FIELD SAMPLING

Soil Biochemistry, Volume 1, edited by A. D. McLaren and G. H. Peterson
Soil Biochemistry, Volume 2, edited by A. D. McLaren and J. Skujiņš
Soil Biochemistry, Volume 3, edited by E. A. Paul and A. D. McLaren
Soil Biochemistry, Volume 4, edited by E. A. Paul and A. D. McLaren
Soil Biochemistry, Volume 5, edited by E. A. Paul and J. N. Ladd
Soil Biochemistry, Volume 6, edited by Jean-Marc Bollag and G. Stotzky
Soil Biochemistry, Volume 7, edited by G. Stotzky and Jean-Marc Bollag
Soil Biochemistry, Volume 8, edited by Jean-Marc Bollag and G. Stotzky
Soil Biochemistry, Volume 9, edited by G. Stotzky and Jean-Marc Bollag
Soil Biochemistry, Volume 10, edited by Jean-Marc Bollag and G. Stotzky

Organic Chemicals in the Soil Environment, Volumes 1 and 2, edited by C. A. I. Goring and J. W. Hamaker
Humic Substances in the Environment, M. Schnitzer and S. U. Khan
Microbial Life in the Soil: An Introduction, T. Hattori
Principles of Soil Chemistry, Kim H. Tan
Soil Analysis: Instrumental Techniques and Related Procedures, edited by Keith A. Smith
Soil Reclamation Processes: Microbiological Analyses and Applications, edited by Robert L. Tate III and Donald A. Klein
Symbiotic Nitrogen Fixation Technology, edited by Gerald H. Elkan

Sustainable Agriculture and the International Rice–Wheat System, edited by Rattan Lal, Peter R. Hobbs, Norman Uphoff, and David O. Hansen

Additional Volumes in Preparation

Plant Toxicology: Fourth Edition, Revised and Expanded, edited by Bertold Hock and Erich F. Elstner

FIELD SAMPLING

Principles and Practices in Environmental Analysis

ALFRED R. CONKLIN, JR.
Wilmington College
Wilmington, Ohio, U.S.A.

with
ROLF MEINHOLTZ
Environics USA
Port Orange, Florida, U.S.A.

MARCEL DEKKER, INC.   NEW YORK · BASEL

Library of Congress Cataloging-in-Publication Data
A catalog record for this book is available from the Library of Congress.

ISBN: 0-8247-5471-9

This book is printed on acid-free paper.

Headquarters
Marcel Dekker, Inc.,270 Madison Avenue, New York, NY 10016, U.S.A.
tel: 212-696-9000; fax: 212-685-4540

Distribution and Customer Service
Marcel Dekker, Inc.,Cimarron Road, Monticello, New York 12701, U.S.A.
tel: 800-228-1160; fax: 845-796-1772

Eastern Hemisphere Distribution
Marcel Dekker AG,Hutgasse 4, Postfach 812, CH-4001 Basel, Switzerland
tel: 41-61-260-6300; fax: 41-61-260-6333

World Wide Web
http://www.dekker.com

The publisher offers discounts on this book when ordered in bulk quantities. For more information, write to Special Sales/Professional Marketing at the headquarters address above.

Current printing (last digit):

10 9 8 7 6 5 4 3 2 1

PRINTED IN THE UNITED STATES OF AMERICA

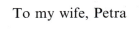

To my wife, Petra

Preface

Field Sampling: Principles and Practices in Environmental Analysis is based on more than 30 years of personal experience in environmental sampling and analysis. This work involved walking fields, taking samples, putting them in appropriate containers, and keeping a record in a project notebook. Samples were brought back to the laboratory where I made the solutions and extractions necessary for the analyses and then carried them out. This includes both simple analyses, such as pH and titrations, and complex analyses involving gas chromatography, HPLC (high-precision liquid chromatography), infrared spectroscopy, atomic absorption, and flame emission spectroscopy.

Rolf Meinholtz, who authored Chapter 10 and coauthored Chapter 11, has more than a decade of experience analyzing environmental samples. This work has included sample extraction and preparation, particularly for but not limited to volatile compounds, and he has completed the actual analyses. He is familiar with and has carried out many U.S. Environmental Protection Agency analytical procedures. He has also maintained and repaired instrumentation used in these analyses.

The basic concept behind this book is to bring together in one place all the information and tools important for successful sampling of the environment and to present it in a manner useful in both commercial and academic settings. To accomplish this, a broad range of topics are introduced. Because some sampling situations require additional investigation of topics (e.g., ground penetrating radar, global positioning, etc.), references have been provided that allow for this. In no case is this book intended to reflect, convey, or impart knowledge, policy, or legal information about any particular type of sampling, restrictions, or compliance codes relevant to any governmental, regulatory, or legal organization or institution.

The tendency of many investigators is to go to a contaminated field and immediately start taking samples; however, this approach is wasteful of natural resources. Making the best sampling plan and using it saves time, limits human exposure to toxic situations, and conserves chemicals and instrumentation. Even though it may initially seem to be time-consuming and expensive, in the long run a good presampling and sampling plan saves money. Proper application of modeling, statistics, and analytical methods also saves manpower, limits exposure of workers, and saves money.

Analytical methods of analysis are accurate, precise, and very sensitive. It is possible in some cases to be able to look at single atoms and molecules. However, soil, regolith, and environmental samples in general show high variability. Soil, regolith, decomposed rock, water, and air are constantly moving. Material is being added and lost on a minute-to-minute basis. A sample is a very small portion of this moving environment, and because of this there is significant variability in all environmental samples.

In addition to the variability inherent in the material being sampled, there is variability introduced by the acts of sampling, handling, and storage of samples before analysis. Of these three the actual process of deciding what, where, and how to sample is the greatest source of error. Because of this it is essential to have detailed knowledge of the material being sampled. What are its characteristics, what should we expect it to contain, and what kind of variability is inherent in the medium? Knowing the answers to these types of questions will mean that we make better decisions in developing and executing a sampling plan.

Many think of field sampling as simply going out to the field, getting some material, and analyzing it. However, this is not an effective way to ascertain what is present. Today there are tools, beyond just going to the field and sampling, that can greatly enhance the sampling process. GPS (global positioning system) can be used to locate the contaminated field and the limits of the contamination therein. In addition, it can be used to identify sampling sites so that they can reliably be sampled at a later time. Some would argue that flags or other markers can be left at sampling sites to facilitate repeated sampling at the same place. Unfortunately, all too often such markers are lost, covered, or moved without the investigator's knowledge. Expensive GPS instruments are available but so are simple, inexpensive systems, and often there is little difference in accuracy and precision between expensive and inexpensive systems.

GIS (global information systems) can be used to locate pollution and its extent and to tell the user whether the field is receiving additional contamination from adjacent areas. It can also be used to follow the progress of remediation. This is accomplished by producing maps with overlays of the contaminated field and surrounding areas. Because GPS, GIS, and GPR (ground penetrating radar) provide additional information about an affected field and its characteristics, they can be important tools in obtaining a better, more representative sample for analysis. They also allow for mapping and provide valuable data for modeling.

Digging is one way to find out what is below the surface; however, GPR can also be used to find out what is below the surface. Using this technology can alleviate or lessen the necessity for extensive sampling or movement of contaminated soil, sediment, or regolith. It can also eliminate the possibility of dinging into gas and water pipes and data, phone, and electric lines and possibly explosives.

Modeling programs can be used to predict where contamination is and where it is likely to go in the environment, and thus to guide sampling. The data can then be used to prevent the spread of contamination and contain the cost of cleanup or remediation. Modeling programs, although useful, should always be checked against field evaluations including laboratory analytical results.

Safety is extremely important when approaching an unknown field even if the field appears open and without hazards. A safety plan needs to be prepared and implemented before sampling begins.

In Chapter 10 some basic information about analytical methods is given. This may seem out of place in a sampling book. However, what happens in the field and what happens in the laboratory are interconnected. Ultimately, what happens in the field affects both the precision and accuracy of analytical results in the laboratory. Laboratory results can indicate that something has changed in the field and needs to be addressed. This is well illustrated in Chapter 11, which discusses traps, mistakes, and errors encountered in sampling. Taking samples from the wrong location, not taking the sample at the correct depth, and handling or storing the sample incorrectly are a few examples.

This book includes all elements one needs to know about in order to take samples that give a representative, accurate measurement of the conditions and concentrations of components of concern in a field. Faithfully following the procedures outlined will provide an excellent picture of a sampled area and its needs, if any, in terms of remediation.

In the references at the end of each chapter are citations for Web sites. These are extremely important because the information they contain tends to be more current than is much 'hard copy" information. Also, some information is available only on the Internet. All the Web references have been checked several times; however, the Web is much more fluid than printed work, which means that sites change as do their addresses. If an address does not work, try accessing the parent site or home page. If an address has 'epa" in it, for example, it is an Environmental Protection Administration (EPA) site and if the full address does not work, go to the EPA home page and look for the material from that location. Most home pages have search engines that will help in locating the needed information.

I wish to thank Rolf Meinholtz for contributing Chapter 10 and parts of Chapter 11 and for offering many suggestions for other chapters. I also thank the following colleagues who helped in reviewing this book: Mon Yee, Paul Thomas, Jennifer Krueger, Sally Ullom, John Ryan, Pam Miltenberger, A. Warrick, Tom Hobbson, H. Bohn, F. Anliot, M. Anderson, and D. Watts.

Remember that local rules apply. Each local area, township, county, state, and country has its own rules for safety, sampling, and approved analytical procedures, as well as individual interpretations and applications of analytical results to remediation of environmental problems. For this reason the information presented is general, and some parts may not be applicable in certain situations and places. Always find out about local rules, interpretations, and applications.

Mention in this book of any program, organization, or company does not constitute an endorsement or recommendation but is simply given as an example of the types of support available to persons involved in field sampling.

Alfred R. Conklin, Jr.

Contents

ix

1

Introduction to Field Sampling

Often there is a great desire to start sampling a field without knowing anything about it or its surroundings and without a well-developed plan. Such an approach will not lead to success in understanding the levels of the component or components of interest contained therein. Successful sampling depends on a careful, thoughtful understanding of the characteristics of the field, the components it contains, the surrounding fields or areas, and its history. In addition, sample handling, transport, and the analytical methods to be used will be essential components of a successful sampling plan.

Depending on where you are in the United States or elsewhere, a field may be a large or small area. It may have an area of thousands of hectares or less than one. In urban areas or in industrial settings a field may be referred to as a site, and the two terms may be used interchangeably. In some cases the two may be put together and termed a field site. The term site is often used in two different ways or senses—that is, to refer to either an area (field) or a precise location within a field. In the latter case it would normally be the specific place from which the sample is taken.

In addition, the term field is used in a generic sense; that is, anytime someone leaves an office, travels to a location, and takes a sample the person is said to be in the field and field sampling. Anytime he or she is out of the office he or she is thus in the field. The sample obtained might be air, water, soil, biological, or any other component of the environment. Note that safety personnel have other definitions for the term field sample. For the U.S. Occupational Safety and Health Administration (OSHA) a field sample may be made for noise or for other safety concerns, such as guards on moving machine parts. Although the emphasis here will be on field as a land area, the methods described apply to field sampling in all respects. A sampling plan, a field sampling project notebook, safety considerations, and so on, will all need to be taken into consideration in all field sampling situations.

In this book we will use these terms in a specific way. Field will refer to any area to be sampled, including fields, water, sediments under water, and air. Site will refer to a specific location in the field from which a sample is to be taken. Figure 1.1 shows a landscape that would contain many fields. It also shows subsurface features that would affect sampling. A field sample may be obtained by physically sampling all or a portion of this landscape.

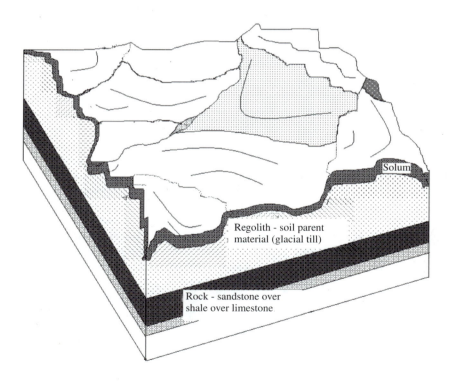

FIGURE 1.1 A landscape including surface and subsurface features.

Sampling can include any method of obtaining information, such as remote sensing, about the condition or components in the field. When discussing field sampling it is important to make sure that all terms are used in an unconfusing way.

1.1. ASSESSING THE SAMPLING NEEDS

An eye-level examination from one position in a field may provide a biased assessment of the sampling needed. It may also lead to unsafe and costly errors in developing the sampling plan.

In addition to viewing the field, in starting a sampling project some premature work is often done. Statistical tools may be applied to the assumed situation. Other times, someone goes to the field without thinking and grabs a sample and has it analyzed. All this activity may have the semblance of being intended to determine the level of a component or components present in a field. Some useful information may be obtained. To obtain accurate and useful information, however, a detailed, well thought out sampling plan based on an understanding of the environment in general and the specific characteristics of the field to be studied must first be developed.

An essential question to ask is what do we wish to accomplish by sampling this field. In order to be able to answer this question we need to know about the environment so that we know we are getting a sample that will give us the answers we need. We walk outdoors, breathe the air, feel the rainfall, and feel our feet on the ground. It is easy to take all this for granted and not consider the various characteristics of these and other components of the environment and how they may affect a field sample.

To obtain the needed information the general and specific characteristics of the field must be known. It is also important to know how the field is related to the rest of the environment. For example, does water move into, on, through, or under the field? Is the area being used (or has it been used in the past) mainly for manufacturing or agriculture? Is the field low-lying and thus receiving inputs from surrounding areas, or is it high and eroding? All possible and probable interactions of the field with the surrounding environment must be taken into consideration.

The specific type or types of materials contained in the field must be known. Without this information samples from a landfill might be mistaken for samples from a cornfield. What is the source of the material? Is it the natural soil or has it been brought in as fill from some other place? Imported material may unknowingly be contaminated and thus add to the sampling problem.

More specifically, what are the characteristics of the medium (for instance, soil) to be sampled? Is it a sandy soil that has little ability to retain water and chemicals, or is it clay, which has high retentive capacity? Is there only one component, or are there several components of interest present? What are the characteristics of the components? Are they soluble or insoluble in water, toxic or carcinogenic, volatile or nonvolatile, flammable or nonflammable? This information is added to the general information to obtain all the information important to the sampling plan.

A visual evaluation of the present condition of the field, obtained by walking around its perimeter, is important in assessing the situation. Is the field bare or planted, or does it have weeds on it? What is its slope? Is there evidence of erosion? What are the surrounding structures like and what is their relationship to the field? What is its physiography? All these are noted and recorded while walking around the field.

As much information as possible about the background or history of the field is obtained next. In how many different ways has it been used in the past? Do any of these past uses impact the sampling plan? Are there likely to be pipes, tanks, or electrical or communication cables buried in the field? All of this type of information needs to be obtained and added to the other information before sampling is begun.

The understanding gained during the above activities provides information about how extensive and intensive the sampling needs to be. If one or more of the contaminants is volatile, soluble, or mobile the extent of the contamination will be greater than it would be for an immobile contaminant. A contaminant soluble in water and carrying a negative charge (anion) would be expected to move further and possibly contaminate water supplies. This calls for a more extensive sampling plan. A low solubility contaminant may remain in a restricted location for long periods of time and not constitute a hazard to surrounding areas, thus calling for less extensive sampling. Putting this all together allows for understanding the sampling requirements necessary for a successful investigation [1].

Along with this information, careful consideration must be made of the safety measures needed during sampling. The largest part of safety is to prevent contamination of workers during the sampling procedure. For example, fields heavily contaminated with petroleum may be flammable. In this case there can be a fire hazard caused by smoking or sparks in the field, and this is a safety concern. It is also important, however, that persons doing the sampling or other persons present not inadvertently remove contamination from the field. In the above case samples taken off the field may be a fire hazard if not stored and transported properly.

Considering what happens to the sample once it is taken, including during transport, is also important. It is possible that the analyte (component or contaminant) of interest easily undergoes rapid changes both in the field and in sample containers. This may dictate the use of a rapid sampling and analysis procedure rather than slower, more accurate, and more precise procedures. A commonly cited example is nitrogen compounds. Under oxidizing conditions and moderate temperatures ammonia is rapidly oxidized to nitrite and nitrate. Under anaerobic conditions nitrate can rapidly be reduced to nitrogen gas. A two-week lag time between collection of a sample containing nitrate and its analysis would mean that the analytical results would not represent the level of nitrate in the field [2].

1.2. SOIL

Soil could be considered the loose material, derived from stone, on the Earth's surface. There are two distinct ways soil is used, however, for agriculture (growing crops and raising animals) and for engineering (supporting structures). There are thus two distinct ways of describing soil, and any persons involved in sampling must be familiar with both.

1.2.1. Soil Scientist Description

Normally a field will contain one or several soils. (See Figure 1.2.) It may also have rock outcrops and water moving through it (or during a flood, over it). Soil will certainly have water moving under it. If it is a fully developed soil, there will be several subsurface horizons. (See Figure 1.3; also Figure 2.4 in Chapter 2). These horizontal layers will have significantly different physical, chemical, and biological characteristics. They develop as a result of the action of the soil-forming factors, time, topography, parent material, biota, and climate, the effects of which extend 1.5 to 2 meters deep. This volume of active soil development is called the solum.

In addition to the surface A horizon there may be several subsurface horizons, the most common being B and C horizons. The A horizon is distinctive in that it contains more organic matter and is darker than the underlying horizons. The B horizons are distinctive in that they contain more clay than over- or underlying horizons and are usually divided into several subdivisions, depending on changes in horizon characteristics. The C horizon is the material from which the soil is forming and is considered the soil parent material. Although the parent material has not been acted upon by the soil-forming factors it may contain horizons that are a result of the original deposition or formation of this material.

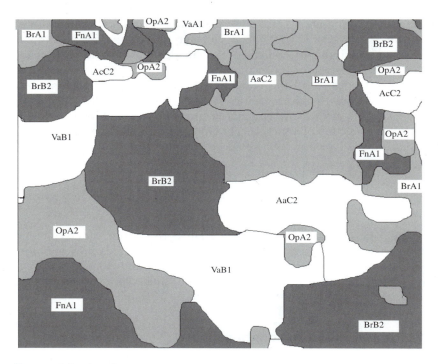

FIGURE 1.2 A soil survey map of several soils showing different soil types with differing slopes and erosion. The first two letters represent a mapping unit (different soil), the following capital letter represents the slope, and the last number represents the amount of erosion.

In the United States soils are classified as belonging to one of twelve soil orders,* which are distinguished by the number and type of horizons they contain. Some orders, such as Gelisols and Aridisols, are also distinguished by their location. Gelisols are frozen soils that occur and develop in frigid regions and contain a permanent frozen layer. All soils in arid or desert regions are called Aridisols. Other soils that have one very distinguishing characteristic are the Andisols, which develop from glassy volcanic ash, and the Vertisols, which develop cracks 30 cm wide and 1 meter deep when dry.

A complete list of the twelve soil orders is given in Table 1.1, along with an indication of a characteristic important in the sampling of each.

---

* Other countries have systems similar to the U.S. system, but often have more orders.

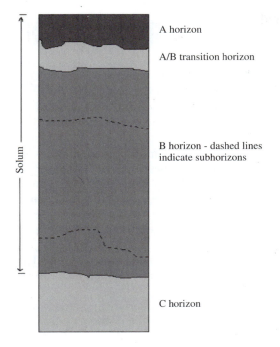

A horizon

A/B transition horizon

B horizon - dashed lines
indicate subhorizons

C horizon

FIGURE 1.3   A simplified soil profile showing the development of several horizons.

TABLE 1.1   The Twelve Soil Orders and Some Important Characteristics

Soil order	Characteristic
Entisol	No horizon development
Inceptisol	Minimal horizon development
Andisol	Developing from volcanic ash
Histosol	Organic soil
Gelisol	Frozen soil
Aridisol	Desert soils
Vertisol	Soils developing large cracks when dry
Alfisol	Horizon development under trees
Mollisol	Horizon development under grass
Ultisol	Horizon development under tropical and subtropical trees
Spodosol	Horizon development in sandy soil under coniferous trees
Oxisol	Most highly weathered

They are given generally in order from least to most developed—that is, from those orders showing the least to the most horizon development [3].

1.2.2. Engineering Description

The engineering description of soils is very different from that of soil science given above. For engineering purposes the carrying capacity, shrink-swell, and plasticity characteristics of soil are more important than its chemical or biological characteristics. The engineering classification gives more emphasis to larger components such as gravel and its characteristics, and less to erosion and crop production characteristics. Engineers describe soils using a two-letter designation that provides an indication of its composition. Thus a soil composed of well-graded gravel would have the designation WG, while a soil composed of clays with high plasticity would be designated as CH.

A soil's plasticity and liquid limit play a large role in its engineering characteristics. The plastic limit is the moisture level when the soil just becomes plastic; that is, it can be molded into a shape and retain it. The liquid limit is the moisture level when the soil becomes liquid, has a loose consistency, and will run.

A more detailed description of soil horizons, underlying materials, and engineering characterizations will be presented in Chapter 2. The importance of these horizons and ways of describing soils in developing a field sampling plan will also be discussed.

1.3. WATER AND AIR

Water and air are completely different from soil, although they make up a significant part of the soil environment. They are more mobile, and contaminants diffuse and are mixed with them more readily. Another difference is that the movement of water and air is associated with currents, which are responsible for the mixing and movement of components (either dissolved or suspended) in them. In the case of air these currents may be called wind. Water and air carry soil and contaminants (both with and without associated suspended particles) for long distances. Eventually materials carried by water and air are deposited either on the soil surface or at the bottoms of bodies of water. The potential and real movement and deposition of soil and other components carried by water and air is extremely important in sampling.

One area in which soil, water, and air are similar is that all three have layers or horizons that have different characteristics. In addition to currents and wind, the layers or horizons from which samples are taken are thus important. However, because of the differences among soil, water, and air, separate specialized sampling tools are used for each. In many cases sample containers for these three will also be different.

In addition to their individual existences, air and water are also integral and active parts of soil. Under idealized conditions a soil sample is half solid and half void space. Under the same conditions soil is envisioned as having half the void space filled with water and the other half filled with air. Figure 1.4 gives a graphical representation of this idealized composition of soil. There is an inverse relationship between soil water and soil air; that is, when it rains the void space is filled with water and there will be little or no air and when the soil dries the space occupied by water is replaced by air. The result is that both soil air and water content are highly variable.

Soil air itself is also highly variable. In general it contains the same constituents as atmospheric air; however, these are in different concentrations in soil. Plant roots and soil microorganisms respire, taking up oxygen and releasing carbon dioxide into soil air. The soil atmosphere is thus higher in carbon dioxide and lower in oxygen than atmospheric air. In addition, it is possible to find highly reduced gaseous species in soil air even when the soil is aerobic. For instance, it is not unusual to find methane in the soil atmosphere along with nitrogen, oxygen, carbon dioxide, and water vapor.

Soil water is variable in concentration, depending on the minerals and amount of water present. In terms of ions the types present are determined by the mineral composition of the soil and so are less variable in any one soil. Soil pH is also determined by the minerals present and the amount of water available for leaching the soil profile. The more water leaching through the soil, the more acidic it will be.

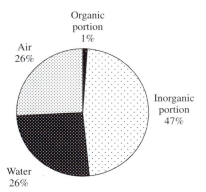

FIGURE 1.4 Composition of an idealized soil sample. Note that the organic composition may be as little as 0.1% for some tropical soils and as high as 30% or higher for histosols (organic soils).

1.4. PRESAMPLING

At the very beginning of any sampling project it is essential to have a project notebook. This will be used to record all information about the project, including the sampling location, the history of the location, and all of the activities in, on, and around it. All entries in the book must be dated, sometimes with the specific time of the activity noted. All samples taken must have on their container a designation, usually a combination of letters and numbers, that directly points back to the pages in the project notebook in which the samples are described.

The description in the project notebook must describe the sample precisely and in depth, including its location and the depth and the date the sample was taken. The amount of sample and any distinguishing characteristics (e.g., gravel, distinguishing color, occurrence or lack of roots, sample taken during a rainy day) are examples of some other important characteristics. Also, the sample container used and any other potentially relevant data available are recorded. (The project notebook will be described in more detail in Chapter 3.)

In addition to the project notebook another presampling activity as noted above is to obtain as complete a history of the location as possible. This information is included in the project notebook. This may seem superfluous at first, but can save a lot of time, energy, and frustration during the sampling process. Uncovering some unforeseen obstacle in the middle of sampling can disrupt the entire program—to such an extent that it may have to be started again from the beginning. (See the example in Section 11.7 in Chapter 11.)

1.4.1. Maps

Several different maps and different types of maps (general area and detailed) are needed for sampling. General area maps are used to determine if adjacent areas can affect or be affected by sampling activities in the field. Detailed maps are used to record sampling activities in the field. For this reason many simple copies of this map will be needed. Whenever an activity is carried out in the field it should be noted on a map. Thus, there should be a separate map for each activity and each day's activities in the field.

An additional essential map is a soils map of the area. Such a map shows the soils and their extent on an aerial photograph. They allow samplers to know which soils are being sampled and to find uncontaminated reference soils to sample. Soils maps, an example of which is shown in Figure 1.2, are included in a soil survey of the area and give invaluable information about the area's characteristics.

In addition to maps, photographs of portions of the Earth's surface are also available. These may be simple pictures or specialized digitalized photographs, such as the digital orthophoto quarter quadrangles (DOQQ). These digitalized photographs are corrected in such a way that they faithfully represent the Earth's surface. Because they are digitalized they can be used with geographic information system (GIS) displays discussed below.

1.5. SAFETY

The safety of the personnel, the field, and the area surrounding the field must always be of concern. Of these three, the safety of the personnel is of prime concern. In field situations unforeseen discoveries are common. They may be simple and interesting or they may be life-threatening. The first rule of field sampling is thus to never sample alone. This is particularly important where the field is hundreds or thousands of hectares in size and will involve being long distances from help, and applies whether the sampling is done on foot, in a boat, on an airplane, or in an all-terrain vehicle. If you cannot walk to safety you may also not be able to row, swim, fly, or drive to safety!

The second rule of sampling is to wear and have available safety equipment. This means wearing appropriate clothing and eye protection at all times. Even under what seem to be the safest conditions additional safety equipment must be readily available.

In addition to keeping the personnel who are doing the sampling safe, precautions must be taken to make sure that unsafe conditions do not develop in other areas either inside or outside the field being sampled [4]. The field office, associated laboratory, and sample storage area are part of this concern. Also of concern is that samples do not spread contamination during storage and transport. More detailed information about safe sampling will be given in Chapter 4.

1.6. SAMPLING

The environment is four-dimensional. It has three physical dimensions—length, width, and depth. The dimension of time also has a significant effect on the contaminant, sampling, and analytical results. When taking samples it is important to take all four dimensions into consideration. For instance, it is not sufficient to only take samples from the surface even if there appears to be no way that the contaminant could have moved through the surface and into the lower layers. To know what is happening to the part of the environment being sampled, samples must be taken at various depths and over a period of time.

1.6.1. Grab Samples

When starting to sample, the first thing people want to do is grab a sample and have it analyzed. Indeed, there is a type of sample called a grab sample. The sample selected for grabbing usually looks different from the bulk material, and thus is assumed to be contaminated. Such a sample might just as well look different because it is uncontaminated and thus it is not known what this sample represents or how it should be handled.

There are two further problems with this type of activity. First is a safety issue. It is not a good idea to go onto a contaminated area without the proper safety equipment. Second, by the time more systematic sampling is carried out the location of this sample will be lost, and so the place of this analytical data in connection to other data obtained will be unknown. There will also be a time difference between this sample and subsequent samples, further lessening its value. Any information gained will thus be of little or no value. Grab samples are potentially dangerous and a waste of time and money.

1.6.2. Preliminary Transect Sampling

When all preliminary nonsampling work is completed, some organized preliminary transect sampling can be initiated. A transect is a straight line passing through the contaminated or other field of interest. Samples are taken at various sites along the transect, both surface and subsurface, and include samples assumed to be outside the contaminated area. A second transect may also be needed if the field contains several different soil types or areas. This transect is chosen so that all soil types or areas are sampled.

Preliminary sampling is important because it is the basis for the application of stastical tools and the development of a realistic, honest, and detailed sampling plan. It is also the basis for determining the handling of the samples from acquisition to final analysis. Handling includes the type of sampler used, sample containers, sample labeling, storage, and transportation. It also provides data needed to determine the type of analysis needed.

1.6.3. Systematic Sampling

On the basis of the historical background and transect sampling a detailed systematic sampling plan can be developed and carried out. This sampling plan includes the areas to be sampled and how sampling sites will be determined and designated. The total number of surface and subsurface samples to be taken will also be part of the plan. Questions such as whether the surface and subsurface will be sampled at each site or only one and whether different soils will be sampled differently or only be noted with a unique label will be answered. Over what length of time sampling is to continue and what criteria are to be used in determining when sampling is terminated will also need to be determined. All of these considerations, along with all other available information, are put together to produce the final detailed sampling plan [5].

1.7. STATISTICS

Statistics and statistical analysis are essential tools in any sampling plan. First, statistical analysis of preliminary sampling data can help determine the number of samples that need to be taken to obtain an accurate picture of the amount and extent of contamination. The same statistic is applied when remediation has been effected to be certain that the field is indeed clean [6].

In any sampling some of the samples will give analytical results that the researcher cannot accept as being valid. A nonbias estimation of the validity of these analytical results needs to be made, however. There are statistical tools that can be applied to determine if the analytical results of such a sample are indeed valid or can be discarded. These tools should be applied to both sample analytical results that are too high and those that are too low [7].

There are also statistical methods called geostatistics, which can be used to estimate the extent of contamination. These tools allow one to map the concentration or level of contaminants over an area. Geostatistical methods can be used in conjunction with global positioning system (GPS) and GIS in developing an effective sampling plan [8].

1.8. MODELING

Sampling, statistics, GPS, and GIS (see below) are all ways of obtaining information useful in making a model of a portion of the Earth. In this case a model that will lead to the most effective sampling plan is produced. Here the word effective includes labor, containment of the spread of contamination, safety, resources, and money.

Models can be physical models of the environment, simple mathematical models of single components, or dynamic, complex models of environmental systems. In Chapter 2 some physical nonmathematical methods for modeling the environment will be described. Although they are nonmathematical, they can lead to an improved and increased understanding of what happens in certain environmental conditions. Models may not be exact, and many refinements may be suggested; however, they do provide valuable and insightful information about the portion of the environment modeled. For instance, a model of the way water moves through porous media of different textures and pore sizes can be informative in many sampling situations [9,10].

There are a number of computer software programs and complex mathematical equations that are useful in modeling water movement through the environment. They can also be used to estimate the fate of contaminants or pollutants. Many of these programs are specifically aimed at predicting the movement of water through porous media, including soil, rock, and fractured rock. Such models can be invaluable in developing an effective sampling and remediation plan. Models are only as good as the data used in them, however. Of particular concern are areas in which water moves through layers of different media or rock that have significantly different hydraulic conductivity (K) and effective porosities (theta; θ). It is also good to keep in mind that the contaminant may affect both K and theta by changing the viscosity of water, which in turn changes the way it moves through the environment. Modeling is treated further in Chapter 7.

1.9. SAMPLE HANDLING

Once the sample is obtained it will be analyzed. Some analytical data, such as pH, will be obtained in the field, while other needed data will require sophisticated laboratory analysis. This means that samples will need to be placed in suitable containers and labeled. Labeling is critically important because a sample that arrives at the analytical laboratory without a label is useless. This means the time and effort used in obtaining and transporting the sample is lost. A more costly result is that the sample usually cannot be replaced and thus the data it contained are lost; that is, by the time the problem is found some time has elapsed and so a subsequent sample will not be the same as the previously taken sample. Sometimes this may not be a problem, but in other cases significant changes in the sampling area may have occurred. For instance, a heavy rainfall might change the concentration or distribution of the contaminant.

Depending on the type of sample, the type of component, the analytical laboratory specified, sample handling, storage, and transport procedures must be followed. Some samples may be dried, some must be sealed in their original condition, and some must be kept at a specific temperature between sampling and analysis. During these operations the sample must be safely handled, stored, and transported in such a manner that the constituent or contaminant of interest is not lost or its concentration changed. These conditions must be not only followed, but also recorded and verified using chain of custody forms.

Once samples are in the analytical laboratory they will be handled in an appropriate manner for the contaminant of interest. It is important for the people in charge of and actually doing the sampling to have some idea of what happens to samples in the analytical laboratory, because what happens there can affect what happens both in the field and during sampling (and vice versa) [11]. Specific examples of this type of interaction will be presented in Chapter 11.

1.10. WHAT IS PRESENT?

In any sampling it is always good to know what is present or likely to be present in the sample. This is important for three reasons. The components present or likely to be present will determine what sampler and what sample container is used. They will also determine what safety precautions need to be taken in both the field and the laboratory. For the laboratory it is also good to know the likely concentration of the contaminants, because some analytical procedures may be sensitive to components other than those of primary interest present in the sample. This could mean over- or underestimation of the level of contamination of interest, which would in turn affect the sampling and remediation plan.

Part of this information will be obtained by a thorough investigation of the history of the field. For instance, the types and concentration of components will be different if the field is or has been used for agriculture, manufacturing, housing, or a municipal dump. The length of time a field is used for a particular activity will often indicate the level of contamination likely to be found.

The occurrence of either an inorganic and organic toxic component in a soil sample, however, is not necessarily an indication of contamination. All soils naturally contain low levels of compounds that are toxic in high concentration. For example, all soil contains low levels of such things as lead and arsenic. The analysis of humic materials may result in finding many simple organic compounds, including phenols, alcohols, acids, and aromatic compounds. Thus in a high-sensitivity gas chromatographic/mass spectral analysis of soil or water extracts a low level of a wide variety of organic compounds should be expected to be found [12]. When evaluating the results of soil analysis it is critical to look at the occurrence of toxic compounds, their concentrations, and the natural occurrence of these same compounds in similar but uncontaminated soils.

1.11. ANALYTICAL METHODS

Samples will be analyzed in several different ways using three different methodologies. Some samples (e.g., transect samples) will be analyzed for some components using rapid methods designed to determine the extent, width, breadth, and depth of contamination. These analyses may be done in the field as the samples are taken or at the field office laboratory. Samples from detailed sampling will be analyzed for some characteristics in the field or at the field office laboratory, but most analysis will be done at a commercial laboratory.

Those samples sent to the commercial laboratory will be treated in a manner appropriate for the contaminant or component that the organization submitting the sample wishes to know about. This is important because samples to be analyzed for metals will be treated differently from samples to be analyzed for organic contaminants, and the two procedures are not compatible. A sample extracted for metals cannot subsequently be used for the analysis of organic contaminants and vice versa.

It is advisable to use the same procedures, analytical methods, and analytical laboratory for all samples. For example, it is not advisable to change the method of determining soil pH in the middle of the sampling plan. For some extraction procedures slight changes in the way the procedure is carried out can make a significant difference in the analytical results. The results will be more consistent, and it will be easier to interpret them if they are all carried out the same way and are sent to the same laboratory [13].

1.12. TRAPS

There are many traps into which the inexperienced or unknowledgeable person doing sampling can stumble in the field. The scenario below is an example of what can happen.

Scenario

An agricultural field planted to sugarcane and having a 5% slope is sampled by two people. No instructions are given to the persons doing the sampling. The first person's samples are found to have phosphate levels equivalent to 200 kg per ha, and remediation is recommended to prevent contamination of adjacent water supplies. The second person's samples have phosphate levels equivalent to only 50 kg per ha, and no remediation is recommended.

From this scenario the question arises as to how such different results could be obtained from the same field. The answer is both in the depth to which the soil was sampled and the place from which the sample was taken. A sample of the top 4 cm of a soil in a depressional area will usually show a high level of whatever you are looking for. On the other hand, a sample of the top 30 cm of soil at the highest point in the field will show a lower level of what you are looking for. The first step in solving this riddle is to make sure that all samples are taken to the same depth. The second is to make sure that random samples are taken from the whole field. A third point is to know the history of the site to make sure that previous usage is not contributing to the differences in the results. The fourth point is to know where phosphate fertilizer was applied and account for this during sampling.

When thought about, this makes sense. Contaminants, fertilizer elements, pesticides, and toxic substances will erode from the high places in the field and collect in the low areas. In the middle between the two extremes the results will be closer to the average. Samples taken from throughout the field are required in order to have a representative indication of the level of phosphate, in this case, in the soil.

1.13. ESSENTIAL UNITS

A common unit used to express the concentration of components or contaminants in the environment is parts per million (ppm). This is a base unit of measurement and can be thought of as 1 part per 1,000,000 parts. Usually we use grams or kilograms, so 1 g in 1,000,000 g would be 1 ppm. It may also be expressed as $\mu g/L$, which assumes that 1 ml of water weights 1 g. It is then valid, because a liter is equal to 1,000,000 μL.

Today you will also find units such as ppb and ppt (parts per billion and parts per trillion) being used. (See Table 1.2.) The amount of a component or contaminant in the field is easily calculated using these units

TABLE 1.2 Common Units Used to Express the Concentration of a Component in a Field

Prefix	Value	Abbreviation
Micro	0.000001[a]	ppm
Nano	0.000000001	ppb
Pico	0.000000000001	ppt
Fento	0.000000000000001	ppq
Atto	0.000000000000000001	ppa

[a]Equals 1/1,000,000.

when used in conjunction with a media's bulk density. These units also allow for easy conversion between the unit, the mass of material, often soil, and the amount of component or contaminant present. Other units of importance include bulk density and what is called a hectare furrow slice.

1.13.1. Bulk Density and Particle Density

In field sampling a sampler is inserted into the soil and a sample extracted. When using a core or bucket sampler a volume of soil is obtained. To understand the relationship between sample volume and the mass of soil it is essential to know the bulk density of the soil. As noted above, soil is composed of both solids and void space filled with water and air. Because of this the term bulk density is used rather than density.

The bulk density of a soil is the amount of soil per unit volume. The official units for bulk density are Mg/m^3. However, the older units of g/cc will commonly be encountered. Both units are the same in that $1.24\,Mg/m^3 = 1.24\,g/cc$ ($1\,g/cm^3$). Most soils have bulk densities between 1.00 and $1.70\,Mg/m^3$. Lower or higher bulk densities indicate an unusual situation, such as an organic soil or a compacted layer.

Knowing the bulk density the mass of soil taken can be calculated if its volume is known. On the other hand, knowing the mass the volume can be calculated. The bulk density will disclose the occurrence of compacted zones in soil, which can affect sampling. Often it is also important to know how much of the sample is solid and how much is void space. For this determination the particle density of the soil particles is needed.

Soil scientists assume that the individual particles in soil have an average density of $2.65\,Mg/m^3$. From this it is obvious that soil must contain a significant amount of void volume. As noted above, a soil sample is assumed to be half solid and half void space. The void space is assumed to contain water and soil air. The percentage of the soil sample with this void space can be calculated by dividing the bulk density by the particle density. More information about the calculation of void space is given in Chapter 2. Other more specific examples of how bulk density is used in sampling will be given in later chapters [14].

1.13.2. Hectare Furrow Slice

Fields are measured in square units. In most of the world the units used are hectares (ha $10^4\,m^2$). In the United States and some other countries acres are used (one acre equals 0.405 ha). Because farmers often work the soil 15 to 30 cm deep we talk of a hectare furrow slice (hafs). (A furrow is a cut, usually between 15 and 20 cm deep, made in the soil by a plow.) If a soil has a bulk density between $1.3\,Mg/m^3$ and $1.4\,Mg/m^3$ and the soil is worked to a depth of 15 cm, an hafs will weigh approximately 2,000,000 kg. This same type of calculation done with acres gives a value of approximately 2,000,000 lb per acre furrow slice (afs). All of these units will be used extensively in this book.

1.14. DEFINITIONS

Environmental work and field sampling will require knowledge of a number of abbreviations and acronyms. These terms abound in environmental work and literature, especially in field sampling and environmental cleanup. The U.S. Environmental Protection Agency (USEPA) has a Web site called "Terms of Environment" devoted to such terms. The section of abbreviations and acronyms beginning with A through D is 13 pages long, and thus the complete list will not be given here. Table 1.3, however, gives a very few commonly encountered abbreviations and their explanation. There has been no attempt to try to pick only the most commonly used abbreviations for inclusion in Table 1.3. This Web site is especially helpful when trying to understand environmental reports [15].

Because many concepts, procedures, and terms are known by their acronyms and abbreviations, many are included in the following chapters. Sometimes it is hard to find information about these concepts, procedures, and terms if their acronyms are not known. A list of acronyms and abbreviations used in this book are included in Appendix A.

Several of the abbreviations given in Table 1.3 are very commonly used. VOC, which stands for volatile organic compound, is an excellent example of one of these. VOCs are compounds having low boiling points, below 200 °C, or high vapor pressure. Common solvents such as acetone,

TABLE 1.3 Environmental Abbreviations and Acronyms and Terms (ATT)

ATTs	Explanation
BMP	Best management practice(s)
CFC	Chlorofluorocarbon
DO	Dissolved oxygen
GEMS	Global environmental monitoring system; global exposure modeling system
MDL	Minimum detection limit
O_3	Ozone
PM10	Particulate matter 10 μm or smaller
RCRA	Resource Conservation and Recovery Act
TCP	Transport control plan
TPH	Total petroleum hydrocarbon
VOC	Volatile organic compound

methylethylketone, diethyl ether, which is commonly just called ether, and hexanes are examples of compounds that fall in this category. Total petroleum hydrocarbons (TPHs) are any hydrocarbons from a petroleum source and commonly refer to fuels such as gasoline, kerosene, and diesel fuel. Another common reference is to heavy metals, which are any of the metallic elements from the lower part of the periodic chart [16].

A very important term that is not an abbreviation is *hot*. Normally we understand this to mean that something is at a high temperature. In sampling we understand this to mean that a sample has a high concentration of contaminant. Any sample with the word hot on it thus should be handled especially carefully. Having *hot* on a label will also alert laboratory personnel to the high concentrations. This can mean that the sample is handled and analyzed in such a way as to obtain the best possible results.

1.15. OTHER WAYS OF SAMPLING AND REPRESENTING THE ENVIRONMENT

There are many other ways of obtaining information about a field. Two of the most important are ground-penetrating radar (GPR) and remote sensing. Ground-penetrating radar might be considered remote sensing; however, it is not similar to most remote sensing in that it involves moving a radar system over the soil surface. Remote sensing is generally understood to involve taking or sensing the Earth's surface from an airplane or satellite, but radar used in weather prediction can be considered a common example of remote sensing.

1.15.1. Ground-Penetrating Radar

Ground-penetrating radar is a relatively recent development. An instrument emitting radar waves is moved along the Earth's surface. Microwaves in the range of 80 to 1000 MHz (megahertz) are directed into the soil. Microwaves are reflected by buried objects with differing dielectric or insulating properties, and the returning waves are detected by a receiving antenna and displayed on a monitor. Such an instrument can locate underground features and obstructions that can hinder or prevent sampling and/or remediation. It can also prevent costly and embarrassing mistakes, such as cutting power or communication lines. Likewise it can prevent the breaking of pipelines carrying water, gas, fuel, or other materials. In the latter case it can prevent additional pollution, contamination, or fire and explosion.

There are other methods of detecting buried objects. Any physical parameter that varies with the composition of the soil will give a signal that can be detected and mapped. Resistivity (resistance per unit of area), electromagnetic induction, and magnetic susceptibility are some of the properties that have been used to investigate subsurface features in soil. These methods all have severe limitations that restrict their applicability and usefulness.

In spite of limitations there are occasions in which GPR and these other methods are extremely useful. Buried explosives and explosive devices are likely to be found in any field traversed by an army. These methods can detect both metal and nonmetal explosive devices buried in soil. Detection of these devices is a safety issue, but their occurrence will also alert the persons sampling to other possible field contaminants [17].

1.15.2. Remote Sensing

Remote sensing is another way to obtain information about a field. A sensor placed in the soil and connected to a radio so that sensor readings are sent back to a central location is remote sensing. Sensors held over the surface of a field on arms attached to a truck would also be remote sensing. Often this type of sensing is used to calibrate sensors for specific use in airplanes and satellites. Normally remote sensing is understood to refer to using either airplanes or satellites to obtain information about the condition of the Earth's surface and subsurface. The sensors may be cameras that obtain black and white or colored pictures of the Earth's surface, or they may record electromagnetic radiation in other parts of the spectrum.

More advanced remote sensing involves the use of spectrometers or special films. These are sensors that record the electromagnetic radiation reflected in a narrow range of wavelengths from the Earth's surface. Some sensors record wavelengths in the ultraviolet part of the spectrum, some in part or all of the visible spectrum, and some in the infrared region of the spectrum [18].

1.16. GLOBAL POSITIONING SYSTEM (GPS)

The GPS is useful in finding a position on the Earth's surface. Signals from two or more satellites are combined by a receiving unit to provide a person or vehicle's location, altitude, and speed. Airplanes and ships routinely use this type of information between airports or at sea. It is also available on some automobiles, and there are also handheld units that an individual can use. Receiving units, either on a vehicle or handheld, can be used in sampling to identify sample sites. This capability can be extremely useful, especially during first-time and subsequent sampling of the same field. It is common to mark sampling spots using various types of physical markers.

Unfortunately it is also common for markers to be removed, lost, or destroyed between sampling times. The GPS can guide samplers to the same sampling location even if the marker was forgotten or has been moved or destroyed.

The GPS is also a surveying tool. It can make finding direction and elevation of points far apart much easier and more accurate than can traditional surveying tools. It can make surveying points separated by obstructions possible. The GPS can also be used to follow the spread of pollution or a toxic compound. Not only can the direction of spread be followed, but also its speed. This provides valuable information that can dramatically decrease the deleterious impact of spills such as those that occur when an oil tanker runs aground.

All activities that use GPSs are termed precision. The use of GPS in farming activities would be termed precision farming or precision agriculture. Weather stations having a GPS receiver would be called precision weather stations. Field sampling using GPS would be called precision sampling [19].

1.17. GEOGRAPHICAL INFORMATION SYSTEMS (GIS)

Satellite and any other digital information can be combined on an image of the Earth's surface using a GIS. This is a computer program that allows any digital information about the Earth's surface to be superimposed on a picture of that part of its surface (e.g., temperature and rainfall). In a like manner pollution concentration in a certain location can be mapped on a picture, aerial photograph, or satellite image of that location using a GIS system.

Although aerial photographs and satellite images are of the Earth's surface, it is also possible, using GIS, to map any known information about subsurface features and thus make a three-dimensional map of an area. Using GIS the location and movement of the pollutant underground or in caves can be mapped using images of the Earth's surface as a reference.

Geographical information systems are systems that allow not only the mapping of the Earth's surface and subsurface, but also the relating of the data in these maps to other data. It makes use of GPS, remote sensing, and other means of obtaining data about the environment, including taking physical samples for analysis in the laboratory. All these data can be related to a map of the area of concern and can then be used to analyze all the data related to the area [20].

1.18. CONCLUSIONS

Before thinking about sampling a field or any part of the environment a firm grasp of important characteristics of the environment relevant to sampling must be understood. These characteristics must also be kept in mind while developing the sampling plan, while carrying out the sampling, and during the transport and analysis of the samples. The four most important considerations derive from the fact that the environment is four-dimensional. The area to be sampled has length, width, and depth, and changes with time. Samples must be obtained from various places on the surface and from the subsurface using a well-developed plan taking these into consideration. The characteristics of samples and the constituents they contain will vary over time. It is not acceptable to take samples from part of an area one month and then a month later sample another part of the area and put the analytical results from these two sampling times together and assume or assert that they represent one sampling.

QUESTIONS

1. What must one know about a field before developing a sampling plan?
2. Would the area to be sampled be expected to be larger with a mobile or immobile contaminant or pollutant?
3. Describe the characteristics of an A horizon. What common component in the A horizon might have the greatest effect on the analytical results?
4. Of the four dimensions of a soil, which is uncontrollable or is constantly changing in one direction?
5. What information is usually available from soil maps and soil surveys?
6. Explain what the following abbreviations stand for:
   (a) BMP
   (b) CFC
   (c) MDL
   (d) VOC
   (e) O_3
7. Explain why ppm is such a handy unit when dealing with soil or solum material. If a sample of soil is found to contain 10 ppm naphthalene, how much naphthalene in grams is found in a hectare furrow slice of this soil?
8. What are the units used to express bulk density in soil media? What is an equivalent older unit?
9. GPS and GIS go together. Explain how.

10. What three statistical tools are commonly used in dealing with environmental sampling?
11. Knowing what is likely to be present in an environmental sample can be important in three ways. Explain.

REFERENCES

1. Patterson GT. Site description. In: Carter MR, ed. Soil Sampling and Methods of Analysis. Ann Arbor, MI: Lewis, 1993:1–3.
2. Brady NC, Weil RR. The Nature and Properties of Soils. 12th ed. Englewood Cliffs, NJ: Prentice Hall, 1999:495–511.
3. Soil Survey Staff. Soil Taxonomy: A Basic System of Soil Classification for Making and Interpreting Soil Surveys. Soil Conservation Service, U.S. Department of Agriculture. Agriculture Handbook no. 436. Washington, DC: U.S. Government Printing Office, 1975:13–71.
4. Petersen D. Human Error Reduction and Safety Management. 3rd ed. New York: Van Nostrand Reinhold, 1996.
5. Créptin J, Johnson RL. Soil sampling for environmental assessment. In: Carter MR, ed. Soil Sampling and Methods of Analysis. Ann Arbor, MI: Lewis, 1993:5–18.
6. Kempthorne O, Allmaras RR. Errors and variability of observations. In: Klute A, ed. Methods of Soil Analysis: Part 1—Physical and Mineralogical Methods. Madison, WI: American Society of Agronomy and Soil Science Society of America, 1994:1–31.
7. Dixon WJ. Extraneous values. In: Klute A, ed. Methods of Soil Analysis: Part 1—Physical and Mineralogical Methods. Madison, WI: American Society of Agronomy and Soil Science Society of America, 1994:83–90.
8. Warrick AW, Myers DE, Nielsen DR. Geostatistical methods applied to soil science. In: Klute A, ed. Methods of Soil Analysis: Part 1—Physical and Mineralogical Methods. Madison, WI: American Society of Agronomy and Soil Science Society of America, 1994:53–81.
9. Environmental Modeling with GIS. Goodchild MF, Parks BO, Steyaert LT, eds. New York: Oxford University Press, 1993.
10. Conklin A. Soil Demonstrations for Geoscience, Environmental Science, Soil Science, Environmental Chemistry, Soil Chemistry, Earth Science, Environmental Engineering. Wilmington, OH: Petra Publishing, 2000:137–140.
11. Bates TE. Soil handling and preparation. In: Carter MR, ed. Soil Sampling and Methods of Analysis. Ann Arbor: Lewis, 1993:19–24.
12. Bohn HL, McNeal BL, O'Connor GA. Soil Chemistry. 2nd ed. New York: Wiley, 1979:272–315.
13. Stevens JL, Green NJL, Bowater RJ, Jones KC. Interlaboratory comparison exercise for the analysis of PCDD/Fs in samples of digested sewage sludge. Chemosphere 2001; 45:1139–1150.

14. Blake GR, Hartge KH. Bulk density. In: Klute A, ed. Methods of Soil Analysis: Part 1—Physical and Mineralogical Methods. Madison, WI: American Society of Agronomy and Soil Science Society of America, 1994:363–375.
15. Abbreviations. USEPA. http://www.epa.gov/ search for CEPPO.
16. Hong J. Environmental Acronyms. Rockville, MD: Government Institutes, 1995.
17. Conyers LB, Goodman D. Ground-Penetrating Radar: An Introduction for Archaeologists. New York: Rowman & Littlefield, 1997.
18. Remote Sensing Change Detection: Environmental Monitoring Methods and Applications. Lunetta R, Elvidge C, eds. London: Taylor & Francis, 1999.
19. Letham L. GPS Made Easy. 3rd ed. Seattle: The Mountaineers, 2001.
20. Korte GB. The GIS Book. 5th ed. Albany, NY: OnWord Press, 2001.

2

Characteristics of the Environment

In order to successfully sample a portion of the environment one must understand the characteristics and dynamics of the portion of the environment being sampled. It is also critical that the relationship and interaction between the specific part of the environment being sampled and the rest of the environment be understood. The term successful sample refers to a sample that truthfully and accurately describes the characteristics of the part of the environment sampled at that particular time. In addition to using the results of a sample analysis to predict the future of a particular component or contaminant in an environment, its relationship to all parts of that environment must be understood.

It might be assumed that the lithosphere is inert and impenetrable to water and contaminants. Sandstones can contain large amounts of water, however, which can move through it at significant rates. Even bedrock, such as limestone and shale, can contain numerous cracks that allow ready movement of water and contaminates. It is important to know not only where a component is at a given time, but also where that component may move to in the future and how a particular environment may change a contaminant.

Another example is the atmosphere, which can affect sampling in three ways, particularly in windy conditions. It can remove contaminants from a contaminated field, thus changing their concentration. It may also deposit contaminants, including particulate matter, such as contaminated silt and clay particles, on a field. To a lesser extent it can also move sand-sized particles from one location to another. Thus, wind can recontaminate an area or a field that has been remediated. Wind can carry contaminants long distances, and so its effects can be widespread.

A third example is water, which may be seen as a nuisance when it rains or when a river floods and covers a field that is to be sampled. Rain can leach contaminants deep into a soil or through a soil and into the water table. If this happens, water from wells in and around the field will need to be sampled and analyzed. Flooding can wash away contaminants, spreading the hazard. It may also mix contaminants with more soil or cover a contaminated area with clean soil that must be removed to effect remediation. While the soil is saturated with water it becomes anaerobic and thus changes the chemical and biological condition of the contaminants. For these reasons water is an extremely important factor in sampling. If contaminants are being carried by water, sampling the water itself will become an important sampling activity.

It might be assumed that soil is a homogeneous mixture with a well-defined movement of air, water, and contaminants when water moves through it. Soil has horizons of differing composition, however, which allow differential movement of water downward through the soil. Some horizons may cause water to move horizontally rather than vertically, thus causing contaminants to be found in unexpected places.

2.1. PARTS OF THE ENVIRONMENT

At first glance it seems obvious that there are several parts to the environment. There is the atmosphere, which is the gaseous layer around the Earth. It extends from the Earth's surface to space, which is often accepted to be 80 km (50 miles) above the surface. The hydrosphere includes all bodies of water covering the Earth's surface. It is variable in depth, depending on the underlying lithosphere. The lithosphere is the solid portion of the Earth. It is not only the dry portion of the Earth's surface, which sticks up above oceans, lakes, streams, and rivers, but also the solid portion underlying bodies of water.

Layers of different characteristics are common in all parts of the environment. (See Figure 2.1.) It is not immediately obvious that the atmosphere and hydrosphere are composed of various fluid layers, but they are. On the other hand, examples of layers in the lithosphere are easily observed. Roads that are cut through mountainsides expose observable layers of both rock and soil.

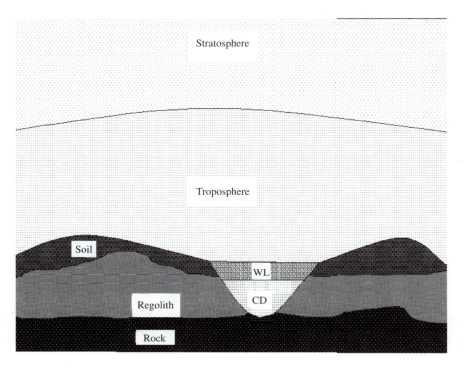

FIGURE 2.1 Layering in the atmosphere, hydrosphere, and lithosphere. CD is cold, light water, while WL is warm, dense water.

2.1.1. The Atmosphere

There are different ways of looking at the layers in the atmosphere. In the older terminology the atmosphere has three layers—the troposphere, stratosphere, and ionosphere. Newer terminology divides it into five layers—the troposphere, stratosphere, mesosphere, thermosphere, and exosphere. This terminology stems from space exploration and extends the atmosphere much further than 80 km out into space. If one looks at the atmosphere from space when it is backlit by the sun, there is a reddish layer near the Earth's surface, a lighter layer above that, and a blue layer above that.

What could the atmosphere possible have to do with field sampling? There are really three possible answers to this question. One has to do with the occurrence and movement of sand, silt, clay, contaminants, and particulate matter in the troposphere. The second has to do with the amount of sunlight, particularly ultraviolet light, reaching the Earth's surface, because some components in the soil may evaporate or vaporize into the atmosphere or be chemically changed by light. The third is that since the atmosphere is part of the environment it may need to be part of the field environment that is sampled [1,2].

Air has a limited carrying capacity for most soil components. Sand cannot be picked up by wind and moved long distances. Silt and clay can, however. Wind can carry atmospheric contaminants long distances from their origin (e.g., the acidification of lakes in the northeastern United States by sulfate released into the atmosphere in the Midwest). During the dust bowl era in the United States some of the dust from the western United States was eventually deposited in the eastern United States and Caribbean. Another well-known example is ash from volcanoes. It can travel great distances around the Earth if it is ejected high enough into the atmosphere. Wind can thus carry contaminants from various sources, and these can be deposited in a field and should be of concern in any field sampling [1,3,4].

Field samples exposed to sunlight for extended periods of time after being obtained from the field can have dramatic changes in their composition. In addition, solar radiation, rain, and wind can also affect the level of contaminant in surface soil through evaporation or vaporization [5,6]. Zimbabwe in south-central Africa is at an average altitude of approximately 1.4 km. It has relatively low rainfall, usually occurring from December through March. During the remainder of the year the sun intensity is quite high and can affect the chemical form of contaminants found in air, water, or soil. Even during the rainy season daytime radiation levels are frequently high. People in Zimbabwe complain about the rapid decomposition of plastics, which is caused by the intense sunlight. Intense solar radiation and its effects on samples are not limited to Africa and Zimbabwe, but can be present anyplace.

The atmosphere will also have indirect effects on field sampling and the resulting samples. These are caused by the composition of the atmosphere and can be both highly variable and dramatic. The atmosphere is composed of nitrogen, oxygen, carbon dioxide, argon, and water vapor plus other gases in low concentration. The nitrogen and argon content of air is constant and inert. The water vapor, carbon dioxide, and oxygen content varies however, and they are reactive. Changes in carbon dioxide will change the pH of associated media, which will change the chemical form and reactivity of contaminants. In the hydrosphere and lithosphere water vapor is less variable, while carbon dioxide and oxygen concentrations are more variable.

On the Earth's surface the composition of atmospheric air may seem unimportant. When taking samples from various depths in water, soil, or sediment, however, the gaseous composition of the material being sampled is important. Samples from submerged or water-saturated zones will have reduced chemical species, which will oxidize when exposed to atmospheric oxygen. This means that if the samples are exposed to air the species identified during analysis will be different from the species actually present in the environment sampled. Different species have different physical, chemical, and biological properties, and thus knowing their exact form is important. Ammonia is a reduced form of nitrogen and is basic. Nitrite is a partially oxidized form of nitrogen and is toxic, and nitrate is fully oxidized nitrogen and is less toxic than nitrite. A sample from a reducing environment can have ammonia in it. When this sample is exposed to air the ammonia can be oxidized to nitrite and nitrate, which can subsequently be reduced to nitrogen gas if the environment becomes reducing again [7].

2.1.1.1. The Soil Atmosphere

At first glance the soil atmosphere is much like the rest of the atmosphere, and it is, in that it contains nitrogen, oxygen, argon, carbon dioxide, and water vapor. Two factors make the soil air different from atmospheric air. First, soil air is constrained in soil pores and because pores are torturous it moves slowly between soil and atmosphere. There are three forces that move air from the soil to the atmosphere and vice versa. Two cause a mass flow of air into and out of soil. The other force causing movement is diffusion.

When the soil warms the air in it expands and moves out of the pores. This happens during sunny days in accordance with the gas law $PV = nRT$. Rearranging this becomes $V = nRT/P$. This equation states that as the temperature increases the volume of gas increases. (All other factors nR and P are constant.) During the day when the soil is warmed air thus expands out of the soil and at night when the soil cools its gaseous constituents decrease in volume and air moves back into the soil.

Rain fills soil pores with water and air is forced out. The little oxygen remaining in water-saturated soils is quickly consumed by plant roots and microorganisms. At this point the soil becomes anaerobic and reducing. When the water moves out of the soil either by evaporation or percolation, air replaces it. The oxygen content increases and the soil becomes aerobic.

Soil air is typically higher in carbon dioxide and lower in oxygen than atmospheric air. Plant roots and soil microorganisms respire, taking in oxygen and giving off carbon dioxide. Carbon dioxide builds up to levels that may be 10 times higher than in atmospheric air. Carbon dioxide diffuses from high concentration in soil air to the lower concentration in atmospheric air. Likewise, oxygen diffuses in the opposite direction [8].

The higher carbon dioxide content in soil air is not insignificant. In soils with pH less than 7 carbon dioxide dissolves in soil water and produces carbonic acid. The acid dissolves soil minerals; for example, calcium carbonate. In soils underlain by limestone this leads to the development of a karst landscape (see Figure 2.2) [9]. In basic soils carbon dioxide reacts with calcium, forming calcium carbonate or lime. Calcium carbonate precipitates out of solution, forming compacted layers that inhibit the movement of water in soils. Heavy metal contaminants may also be precipitated as carbonates and thus change their availability during analysis.

Changes in the oxygen content of soil affect both the chemistry and biology therein. Changes in the anaerobic/aerobic and oxidizing/reducing conditions of a soil sample affect the oxidation states of the metal cations it contains, the oxidation state of nitrogen compounds, and whether or not organic matter is decomposed by aerobic or anaerobic mechanisms. It also affects the primary types of organisms living in the sample. These organisms, particularly microorganisms, are responsible for many chemical changes in the environment. For this reason changes in the organisms present in a sample can dramatically affect the results of an analytical procedure.

The atmosphere can affect the field sampling in many ways and should always be kept in mind. Samples that are oxidizing must be kept that way. Likewise, soils under reducing conditions must be kept in a reducing environment [10].

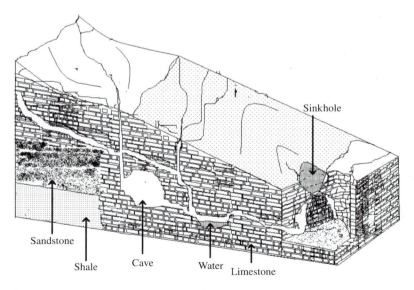

FIGURE 2.2 An example of a karst landscape.

2.1.2. The Hydrosphere

The solution making up the hydrosphere comprises water containing small amounts of dissolved gases and inorganic and organic ions and molecules. Water, with its familiar chemical formula H_2O, is of fundamental importance, both in the environment and in environmental sampling. Water has unique chemical and physical characteristics and chemical reactivity related to hydrogen bonding and attraction to both positive and negative atoms or charges in other molecules. (See Figure 2.3.) It is often called the universal solvent because even the most insoluble components are to a measurable extent soluble in water [2,11].

   Layers of water occur in both oceans and lakes. In northern temperate regions lakes freeze in the winter. During the freezing process the water and the subsequent ice is less dense than the underlying warmer water. During the winter the lower layers gradually cool and become more dense. In the spring the surface melts and the water warms, becoming more dense than the underlying water. At this point the lake turns over, with the upper layer sinking to the bottom and the bottom layer coming to the top.

   Air is more soluble in cold water than in warm water. However, water that has been trapped at the bottom of a lake under ice for an extended period of time becomes anoxic (lacks oxygen). When a lake turns over in the spring, oxygen-rich surface water moves to the bottom of the lake and there is a flurry of oxidative reactions. The composition of the water and sediments in terms of oxidized and reduced species changes dramatically during this time [12].

   In temperate and tropical lakes layers with varying amounts of oxygen are also possible, along with areas of differing temperature. These lakes can also turn over, sometimes with catastrophic effects. Turnover can be due to

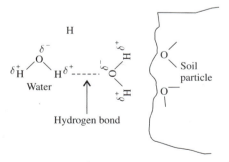

FIGURE 2.3 Water molecules hydrogen bonded to each other and attracted to the oxygens in a soil particle.

incoming streams and rivers, springs, uneven heating caused by currents, and underlying geothermal activity. For these reasons changes in the oxidative state can occur. For areas in which there is a pronounced dry period during the year the bottom of a lake may become anoxic. When the rains come oxygen-rich, water flows into the lake, changing its oxidative status [13].

Even in situations in which turnover or incoming sources of oxygenated water are not of concern, the solubility—even though it is low—of air in water is important. As noted above, air is less soluble in warm water than in cold water, and thus its concentration is dependent on the temperature of the water being sampled. No matter what the temperature is, however, the deeper the water the lower the oxygen content. Thus, at the bottom of bodies of water there is very little (or no) oxygen, producing anaerobic or reducing conditions. Not only are sediments anaerobic and reducing, but so is any portion of the lithosphere saturated with water for a period of time. This condition routinely occurs in soil and in any material covered by floodwaters. Sampling of any water-saturated soils and sediments must thus be done carefully to maintain their physical and chemical integrity.

Soils at the bottom of bodies of water thus need special sampling considerations. They will often need to be sampled, and represent a unique sampling environment. In addition to being anaerobic, soil and sediment at the bottom of bodies of water are under pressure. A field sample taken under these oxygen and pressure conditions must be kept oxygen-free and under pressure until analyzed. Such samples will frequently be easily dispersed and this must be avoided as much as possible. The void spaces between soil particles are completely filled with water, which is called interstitial water. In many cases the interstitial water and the solid particles will need to be analyzed separately.

2.1.2.1. Solutes in Water

Sometimes people are surprised that a chemical analysis of water can result in finding the presence of gasoline components, including ethers. The everyday observation is that these compounds are not soluble in water. There are two answers to this puzzle. First, as already noted, water is capable of dissolving small amounts of almost any material [11]. Second, analytical methods and instrumentation are extremely sensitive. Detection of parts per million was once considered good. Today, however, detection of parts per billion or trillion are routinely and easily made. In some cases it is even possible to look at single molecules or atoms. The detection of a component in water is thus important, but it only tells part of the story.

The occurrence of surfactants (soaps) in water will also dramatically increase the apparent solubility of insoluble compounds. Soaps can come from man-made sources; however, there are a large number of plants that produce soaplike compounds. Both sources can lead to solubilization of insoluble compounds and thus increased levels of these compounds in water.

Suspended, colloidal-sized particles are present in water and carry sorbed material. This material can show up in analysis and be reported in such a way as to indicate that it was in solution. In this way components may appear to be present in water in concentrations above their solubility limit.

2.1.3. The Lithosphere

Geologists classify rocks as being sedimentary, igneous, or metamorphic, depending on how they were formed. Sedimentary rock is the result of rock and rock components being eroded, usually into a body of water. The material sediments and comes under enough pressure to allow the particles to be cemented together. Common sedimentary rocks are shale, sandstone, and limestone. Most often rock is considered hard; however, many sedimentary rocks are quite soft. Some are soft enough to allow them to be broken by hand. Another completely different type of sedimentary rock is coal. This is rock derived from organic matter and is composed of carbon, and thus is very different from other types of rock.

Igneous rock results from the cooling of magma (melted rock material) or lava. This cooling can occur when magma reaches the Earth's surface as lava or when it flows into cracks in nearby rocks and cools. Common igneous rocks are granite, pumice, and basalt.

Metamorphic rock is rock material that has undergone a change in appearance as a result of melting caused by high temperatures and pressure. Sometimes this also causes a change in the mineral composition of the rock. Both sedimentary and igneous rocks can undergo this process. Common metamorphic rocks are slate, marble, and quartzite.

Water can often move through rock, even when it appears solid. Softer rocks will have pores, and when near the surface, even the hardest rock in the Earth's crust will have cracks in it. In both cases pores and cracks allow the movement of water and contaminants downward. There are many springs in the hills on the islands of the Philippines. Rainwater enters the soil and the underlying fractured rock layers on the upper slopes of the hills. This water moves downhill under the ground through the fractured rocks until it finds an outlet. At this point a spring forms, the outflows of which can be large enough to support large waterfalls year round. At other places wells can be dug into these broken rock layers and a good, steady supply of water obtained.

About a fourth of the Earth's surface is underlain by limestone. Two features of limestone are caves and sinkholes. When rainwater enters limestone, soft or easily dissolved portions dissolve, leaving caves, which are sometimes extensive. Such landscapes are called karst, a good example of which is Mammoth Caves in the state of Kentucky in the United States. Limestone caves are common throughout the world, however. When the size of the dissolved area becomes very large the cavity cannot support the roof and the cave collapses inwards. This produces a sinkhole that may or may not have a hole connecting to the cave network (see Figure 2.2) [14].

Water can pass directly from the surface to the underground karst network, which may extend for hundreds of miles, and will have an underground river associated with it. Contamination entering such networks can travel long distances and contaminate large areas. In addition, animals living in the caves, such as bats, can contribute to this contamination. This is particularly dangerous because many water wells may tap into its water system.

2.2. SOIL FORMATION

It is easy to assume that soil is simply decomposed rock lying above a consolidated rock layer. Soil is much more than simply ground-up rock, however. Immediate evidence of this is that ground rock would be gray, while soil is brown to black on the surface with red hues in the subsoil. There are soils with very light colored surface horizons and little color in the subsurface horizons, but even these soils are not gray.

When rock is exposed on the Earth's surface it is subject to degradation. Rain falling on the decomposing rock dissolves salts as it infiltrates. These salts are eluviated with percolating water and moved lower in the regolith. With a moderate passage of time and rainfall soil salts are lost and less soluble components remain, forming soil. Illuviated less soluble materials result in the development of horizons. With sufficient time and rainfall salts and other easily dissolved components are completely leached out of the solum. Most of these salts eventually are deposited in the oceans.

Oceans are not the only reservoirs of salts. When the water falling on mountains is caught in a basin without an outlet to the sea, a salt sea such as the Great Salt Lake in Utah or the Dead Sea is formed. If rainfall is low and evaporation high then a salt flat is formed where there is a crust of salt on the soil surface.

2.2.1. Horizonation

Soils contain horizons, which develop as a result of soil-forming factors. These factors are time, topography, parent material, climate, and biota. The soil-forming factors cannot be put in order of more to less important; all are interdependent and all interact to provide the environment under which the soil is developing. During sampling activities, these factors will be overlain, mixed, and complicated by the activities of plants, animals, and man.

Soil does not develop directly from rock but rather from decomposed rock. The type of soil developed and the time necessary for its development will thus be strongly influenced by the parent material or decomposed rock from which it is developing. If the material is rock ground up during movement by air, water, or ice, the soil will develop faster than if it is developing from slowly decomposing rock.

It takes a long time for a fully developed soil to be formed (see Figure 2.4), perhaps as long as 1000 years. Thus, time is a critical factor in soil formation. In order for time to work the material must be remain in the same place. The tendency to remain or move will depend on the topography of the area, so this is another important factor in soil formation. Soil developing on a hillside will be subjected to more erosion than a soil on a flat plane. A depressional area will have more water and will collect decomposed materials from which its soil is forming. The soil formation will thus tend to be more rapid under these conditions.

Climate is another critical factor during the development of a soil. Salts may be completely leached from soil, while other materials are moved into lower horizons by percolation water. For this reason rainfall is extremely important, and soil will tend to develop faster where there is more rainfall. However, along with rainfall, the average temperature also will be important. The rate of chemical and biological reactions in soil increases twofold for each $10\,°C$ increase in temperature. For this reason the climate under which a soil develops has a strong influence on both its physical and chemical characteristics.

Finally, both the vegetation adapted to the climate and the animals present have a pronounced influence on soil development. Plant roots produce acids and increase the carbon dioxide content in the soil atmosphere. Different soils develop under trees than under grass vegetation because of their different growth and rooting habits. Animals also play a role in soil-forming processes. Worms, termites, moles, groundhogs, and many other animals move soil from lower horizons to the surface. In addition, they leave large holes that water can flow into and through.

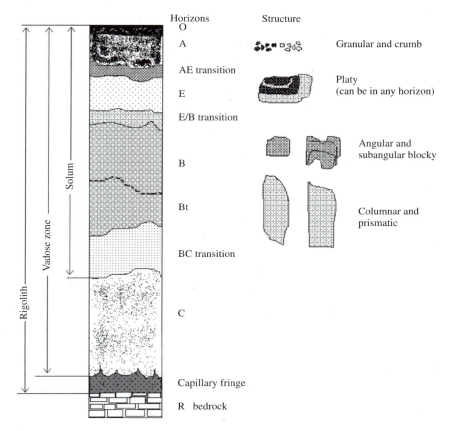

FIGURE 2.4 Designations applied to the layers of loose material over rock. On the right are typical peds found in soil.

All soil-forming—factors, time, topography, climate, biota, and parent material—are interrelated. The amount of rainfall and the average temperature affect the types and abundance of plants growing and animals living in a locality. The amount of water infiltrating the soil and leaching salts out depends both on the rainfall's intensity and amount and on topography, which influences the likelihood of rain infiltrating or running off the surface. All of these are inextricably intertwined with the length of time these factors are at work and the parent material they are working on.

Once a soil has developed, it is characterized and named on the basis of its inherent characteristics. An obvious characteristic is that it has horizons. As in Figure 2.4, the major horizons are designated by a capital letter. The A horizon is the surface horizon usually observable because it has higher organic matter than underlying layers and is thus darker. Under the surface A horizon will occur several other master or major horizons. The most common are the master horizons X, E, B, and C, and they occur in this order from the surface downward. There will also be transition horizons between all the master horizons.

In Figure 2.4 an O horizon is also given. This is an organic horizon, which may be called an Oi, Oe, or Oa, depending on the organic matter decomposition state (Oi represents incompletely, Oe intermediately, and Oa completely decomposed organic material). These horizons are not common, but are important because they are very different from mineral horizons and require different handling, extraction, and analysis.

Horizons designated by two capital letters or a slash between capital letters are transition horizons, depending on the mixing of the two layers. This then indicates that the transition horizons contain characteristics of both the overlying and underlying materials.

In soils developed under tree vegetation an E horizon will be the next major horizon below A. It is lighter in color than the horizons above or below and is coarser. In soils developed under other types of vegetation the A horizon will transition to a B horizon. There will typically be several of these horizons. They are higher in clay and redder than overlying horizons.

The B horizons transition to a C horizon, which is the material from which the soil is developing. It shows no evidence of the effects of the soil-forming factors and may be very thin or thick. Soils developing on underlying rock may have thin, almost nonexistent C horizons, while those developing from glacial till or other deposited materials may have C horizons that are several meters thick. Rock underlying this horizon will often be referred to as the R horizon.

It is common to find situations in which there are more or fewer horizons. A soil may have an A horizon that is underlain by a C horizon. The A horizon may not be present, and the topmost layer may be a B horizon. In cases of severe erosion the topmost horizon may be a C. In river valleys and stream and river terraces there can be buried soils. There may be A and B horizons underlain by another set of A and B horizons. All possible variations of this basic situation can and do occur (e.g., A and B might be underlain by a different B horizon).

Horizons with a capital letter followed by a small letter have distinctive characteristics. In Figure 2.4 the small letter t in Bt indicates that the B horizon has a high amount of clay in it. There are many other ways in which differences in horizons may be designated. These will not be described here; however, if needed they can be found in basic soils or soil taxonomy texts [15]. The local soil conservationist or soil scientist can interpret more complex horizon designations if need be.

Different soils and different horizons may need different sampling. For this reason knowledge of the soils and their characteristics will be needed in developing a field sampling plan.

2.3. SOIL TYPES

On the basis of the characteristics of a soil's profile it is described as belonging to one of twelve soil orders. These range from the Entisols, which have a minimal A (or no) horizon development to Oxisols, which have undergone intensive weathering. A complete list of the twelve soil orders is given in Chapter 1, Table 1.1. Six of these orders, described below, are of particular concern during field sampling.

2.3.1. Gelisol

The Gelisols are soils developing in parent material that is frozen most of the year. They contain permafrost, which is a layer frozen for two or more years in a row within 100 cm of the surface. These soils are slow to develop and may show extreme mixing because of freezing and thawing. Gelisols occur in the northern regions of Russia, Alaska, and Canada. Because of the frozen layer and low temperatures contamination will not be easily removed by natural processes.

2.3.2. Andisols

Andisols develop in volcanic ash and often contain 40–50% cinders, which are larger than 2 mm in diameter. They are young soils that are poorly developed and have high infiltration and percolation rates. Because of their high porosity and lack of profile development pollutants may pass rapidly through them.

2.3.3. Histosols

Histosols are organic soils in the upper 80 cm of the soil profile containing organic matter that can either be fully or partially decomposed. They develop in depressional areas under water-saturated and anaerobic conditions. Organic matter has a high affinity for many soil contaminants, and thus must be treated and analyzed differently than mineral soils.

2.3.4. Vertisols

Vertisols contain large amounts of expanding clays, and because of these clays develop cracks that are 30 cm wide and 1 m deep when dry. When wetted the cracks close. During dry periods surface material falls into the cracks, thus producing an inverted soil. Not only can soil and organic matter fall into these cracks, but also pollutants. This is important to keep in mind when sampling. Remediation of these soils is difficult because water moves through them extremely slowly and the cracks allow pollutants to penetrate deeply into the soil.

2.3.5. Aridisols

Aridisols occur in arid climates, and because there is little rainfall these soils tend to have high pHs, usually above 7.5, and are often affected by salts. This means that pollutants will react differently in these soils when compared to acid soils. The salts will affect remote sensing, particularly ground penetrating radar (GPR), and both the sampling and analysis of samples from these soils.

2.3.6. Spodosols

Spodosols occur in acid, sandy soils and develop under conifer (cone-bearing trees commonly called pine) forests. These soils have a subsurface horizon called a spodic horizon in which aluminum oxides and organic matter have accumulated. Iron oxides may or may not be included in this area. The occurrence of iron and aluminum oxides and organic matter can affect analytical procedures applied to samples from this horizon. This is thus an important horizon that requires special consideration during sampling and analysis.

2.3.7. Other Soil Orders

Each of the other soil orders has its own unique characteristics, but these do not influence sampling procedures as directly as the orders mentioned above. Their various unique horizons, such as the Bt, however, require special attention. Persons carrying out the sampling need to know what soil orders or soil types are present. Local soil scientists can provide needed information about the local soils and their characteristics, which will affect the sampling plan and analytical methods [16].

2.4. SOIL TEXTURE, STRUCTURE, AND BULK DENSITY

Knowledge of soil texture and density is essential in field sampling. Texture can be described using either soil science or engineering descriptions. Density is described as the mass per unit of dry soil, Mg/m^3 or g/cc, and is universally the same.

2.4.1. Texture

The two different methods of describing soil texture depend on whether the soil is to be used for agriculture or engineering. In agriculture it is used for growing plants, and in engineering as a medium to support structures. Soil scientists are most interested in the relative amounts of sand, silt, and clay, as well as the biological characteristics of soil. Engineers are interested in the characteristics of larger stone and gravel, sand, and fine particles. Each of these components is called a soil separate and is defined as having a particular size. (See Table 2.1.) The soil in a field to be sampled may be described by either of these approaches, so it is important to be familiar with both.

2.4.1.1. Texture—Soil Scientist Definition

Soil scientists name soil textures using the percentage of sand, silt, and clay present. Sand is between 2.00 and 0.02 mm in diameter, silt between 0.02 and 0.0002 mm in diameter, and clay less than 0.002 mm in diameter. Using the percentages of sand, silt, and clay and a textural triangle (see Figure 2.5), a

TABLE 2.1 Size Characteristics of Soil Separates Using Soil Science and Engineering Definitions

Soil component	Subdivision	Engineering designation	Soil science designation[a]
Cobbles/gravel		75/4.75	
	Coarse	4.75–2.00	
Sand	Medium	2.00–0.425	2.00–0.02
	Small	0.425–0.075	
Silt		>0.075	0.02–0.002
Clay			>0.002

[a]This is the international definition—the U.S. Department of Agriculture definition is somewhat different.
Note: (All units are in mm.)
Source: Data from Refs. 15 and 18.

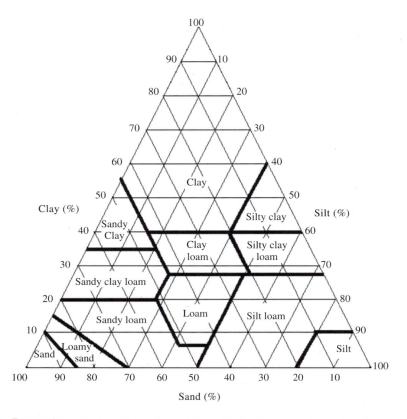

FIGURE 2.5 Textural triangle used by soil scientists to assign textural names to soils of different textures.

textural name of a soil can be determined. A soil containing 10% sand, 15% clay, and 75% silt would have a silt loam texture. The term loam is used for textures for which the sand, silt, and clay fractions contribute equally to the characteristics of the soil. The percentage of clay is lower than silt or sand in loam because it has much higher activity and so it takes less to have an equal effect. Table 2.1 gives the relationship between the soil scientist's and the engineer's definition of soil particle size. If appropriate, a soil's textural name may be prefixed by an indication of larger components. Thus, there might be a gravely silt loam. This would indicate that the soil has a large percentage of gravel-sized material in it. For the soil scientist the particles larger than 2.00 mm in diameter are not as important as the sand, silt, and clay.

In addition to the above names there are other common, nonscientific, less specific terms in common usage, such as light, heavy, clayey, and sandy. Heavy soils are high in clay and light soils high in sand. In regions of the world with a predominance of sandy soils, however, a soil with 1–2% clay might be called clayey. In other areas a soil with 20–30% clay might be called clayey. A similar sort of thing happens with sand. In areas with clayey soils a significant sand content might be called sandy. These common local usages need to be checked by referring to laboratory determinations of texture [17].

2.4.1.2. Texture—Engineering Definition

Engineers classify soil textures using a different classification scheme. They approach the texture of the loose material on the Earth's surface from a broader perspective than does the soil scientist, being concerned with a much larger range of sizes of separates. From an engineering perspective the ability of a material to carry a load is most important. Also from a load-carrying perspective, the uniformity of the material in terms of size is important.

Engineers size soil components by noting the size of the sieve retaining the soil component. The size can then be reported as the sieve number, which is sometimes confusing because as the sieve number increases the size of the holes in the sieve decreases. Table 2.2 gives a number of important sieve numbers and corresponding hole sizes. Material not passing a #4 sieve is cobbles, which are larger than about 75 mm in diameter. On the other hand, gravel is smaller, but is larger than 4.75 mm. Sand is designated as coarse (smaller than 4.75, but larger than 2.00 mm), medium (smaller than 2.00 but larger than 0.425 mm), and fine (between 0.425 and

TABLE 2.2   Sieve Numbers and Hole Sizes

Sieve number	Hole size (mm)[a]
#4[b]	4.75
#10[c]	2.00
#40	0.425
#200	0.075

[a]Actually, the average diameter of the particles of gravel, sand, or silt passing through it.
[b]There are larger hole sieves that do not have the # designation.
[c]This is the upper limit for sand in the soil science textural designations.

0.075 mm in diameter). Material smaller than 0.075 mm is silt or clay. (See Table 2.2.)

Engineers use a combination of capital letters to indicate the texture of the material one is working with. The letters and what they signify are given in Table 2.3. Combinations of letters are used in pairs to indicate a particular type of material. For instance, GW would be well-graded gravels with no fines (silt or clay). On the other hand, CL would be clay with low or slight plasticity. Another possibility would be OH, which is organic matter with high plasticity. In this way a wide variety of materials with varying size components and plastic or organic matter contents can be described.

As can be seen in Table 2.3, plasticity is an important characteristic of soils from an engineering perspective. A plasticity index (PI) can be calculated for any soil or similar material. It is the difference between the plastic limit (PL) and the liquid limit (LL). When water is added to an air dry soil, individual soil peds are moistened first. At a certain water content the soil becomes plastic (PL), which means that it can be molded into a shape and will retain it. If more water is added it becomes liquid and runs; this is the LL. The difference between these two limits is the PI, which is simply calculated using the formula below.

$$PI = LL - PL$$

TABLE 2.3 Engineering Classification of Soils

Symbol	Meaning
GW	Gravel (G) larger than #4 sieve—well (W) sorted
GP	Gravel poorly (P) sorted
GM	Gravel containing silt (M)
GC	Gravel containing clay (C)
SW	Sand (S) smaller than #4 but larger than #200 sieve—well sorted
SP	Sand poorly sorted
SM	Sand containing silt
SC	Sand containing clay
ML	Inorganic silts with low plasticity (L)
CL	Inorganic clay with low plasticity
OL	Organic matter with low plasticity
MH	Inorganic silts of high plasticity (H)
CH	Inorganic clay of high plasticity
OH	Organic matter of high plasticity
Pt	High organic matter soils

Source: Taken from Ref. 18.

Soils with a high PI typically contain large amounts of clay. This gives an indication of the water movement and the retention of contaminants spilled on soil.

There are several other terms or physical characteristics of soils and soillike media that engineers use. One is the Attenburgh limits, which refer to the PL and LL of a particular medium. There is also COLE, which is the coefficient of linear extensibility. As the name implies, this is a measure of the amount of swelling and shrinking a material undergoes between wetting and drying. It indicates the amount and type of clay in a soil. In working with engineers one is likely to hear these terms used frequently.

Engineers will also talk of the *A-line*. This is a graph that relates the PI to the LL. Soils with characteristics above the A-line and with a LL greater than 50 are inorganic clays of high plasticity. In the same region but below the A-line are fine sands or silts and elastic silts. Below the A-line and less than a 50 LL are inorganic silts and very fine sands with low plasticity. In the same region but above the A-line are inorganic clays of low to medium plasticity. (See Figure 2.6.)

When developing a sampling plan the soils in the area may be described by either soil science or engineering terminology. It is important to be able to understand both types of descriptions and use these to determine how sampling should be done. In most cases both descriptions will be valuable. In situations in which gravel is prevalent the engineering

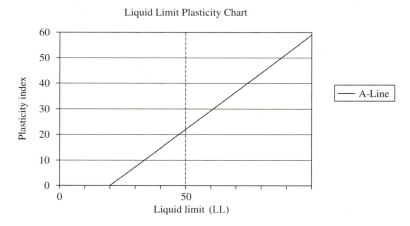

FIGURE 2.6 The A-line used by engineers to describe fine soil fractions. Data taken from Ref. 14.

descriptions will be very useful in deciding on the type of sampler to be used [18].

2.4.2. Structure in Soil

Figure 2.4 shows the components in an idealized, well-developed soil profile. All of the loose material—including water-saturated layers—above rock is called the regolith. The material above the saturated zone is called the vadose zone. The material in which active soil development is taking place is called the solum. The characteristics of these areas is determined by both texture and structure.

The texture of soil is extremely important because it controls important characteristics related to the movement of water and the sorption and retention of contaminants. Texture is not, however, the only factor in movement in the soil. Soil particles, sand, silt, clay, and organic matter do not act independently of each other, but are cemented together to form secondary particles called peds, which are the soil structure. The cementing agents are organic matter, clay, microbial gums, and various cations.

Different sizes and shapes of peds are found in different soil horizons. The different kinds of peds and their typical location in a soil profile are given in Figure 2.4. Note that platy structure or peds can be found in any horizon, but are commonly found between the A and underlying B horizons. Good water-stable peds result in a soil having increased percolation and aeration rates, making it easier to work with and easier to sample and remediate [19].

The increased rates of percolation and aeration are a result of increased porosity in the soil. Pores occur both within peds and within the lines of weakness between peds. Some pores are large and drain readily, while other smaller pores retain water, which is used by plants. Still smaller pores remain filled with water even under the driest soil conditions. More information about pores and their effect on water movement is given in Chapter 7.

2.4.3. Bulk Density

Both soil texture and structure are related to a soil's bulk density. Because bulk soil is composed of both solid and void space or pores it has a variable density, which is specifically called its bulk density. Bulk density is the dry mass of oven-dried soil divided by its volume. Typically the density is obtained by inserting a ring of known volume into the soil using an instrument that does not cause compaction of the sample being taken. The ring is removed and the soil in it leveled. The ring and the soil are placed in a oven at 105° for 24 hr.

The ring plus soil is weighed again, and the weight of the ring is subtracted from the total to give the soil weight. (Equations 2A and 2B in Figure 2.7 are used for these calculations.) At this point one need only divide the mass by the volume of the ring to obtain the bulk density. (Equation 2C is used for this calculation.) Bulk density is usually obtained as grams per cubic centimeter. Soil bulk density is most often reported as Mg/m^3, however.

Typically soil bulk densities range from 1 to $\sim 1.7\,Mg/m^3$. Sandy soils generally have higher and silty and clayey soils lower bulk densities. This is variable, however, depending on the structure and compaction of the sand, silt, and clay. Subsoils have higher bulk densities than surface soils, partially because of lower organic matter content and pressure from overlying soil. Examples of common bulk densities, their associated void space, and associated soil types and conditions are given in Table 2.4.

The void volume in soil is determined using the average particle density of individual soil particles. On the basis of a great many measurements, soil scientists take the particle density of soil to be $2.65\,Mg/m^3$. The amount of solids in a soil sample is calculated by dividing this into the bulk density. Subtracting this from 1 gives the amount of void space. Multiplying these by 100 gives the percentage of each. Equations for these calculations are given in Figure 2.7, Equations 2D and 2E.

Knowing the bulk density of soil in a field we can make many useful and important calculations. Calculations of the kilograms or tons of soil that must be removed or remediated can be determined, as can the volume of soil. Such calculations will also allow mass balance calculations, which allow accounting for all the contaminant present. In field sampling

Volume of sampling ring = Area × Height = $\pi r^2 \times h$ 2A

Weight of soil = oven dry weight sampling ring and soil −
    ring weight   2B

Bulk Density $= \dfrac{\text{oven dry weight of soil sample}}{\text{volume of soil sample}} = \dfrac{g}{cc} = \dfrac{Mg}{m^3}$ 2C

Solid portion $= \dfrac{\text{bulk density}}{\text{particle density}} \times 100 = \%$ solids   2D

% Void space = 100 − % Solids   2E

FIGURE 2.7 Equations for calculating a soil's percentage of solids and void space. In Equation 2A r is the radius of the sampling ring and h is the height.

TABLE 2.4 Common Soil Bulk Densities

Bulk density (Mg/m^3)	Void volume (%)	Soil type	Condition
1 or less	62	Indicates organic soil or soil with high organic matter contact	Light in weight subject to wind and water erosion
1.25	52	Common average bulk density for agricultural soils	Good structure
1–1.6	62–39	Normal range for soils	Good to poor ped structure
1.5 and above	43	Usually clayey soils	Compacted limited water movement through soil
1.7 and above	35	Usually clayey	Compacted

Source: Ref. 17.

situations one may find layers of soil with different bulk densities. The density of the layers must be known if one wishes to obtain comparable samples (on a weight basis). To obtain the same amount of soil solid and the contaminant it contains, it is essential to take different volumes of soils, so that the mass of solid obtained is equivalent. This would mean that the volume of sample would need to be different for the different layers. This type of calculation is illustrated below.

Calculating Volume of Soil Needed for Equal Mass of Samples

A soil with a bulk density of 1.2 Mg/m^3 is to be compared to a soil layer with a bulk density of 1.5 Mg/m^3. The sampler being used has a cross-sectional area of 5 cm². The sampler itself is 30 cm long. How many centimeters of soil should be taken from each layer so that the same amount of soil is obtained for analysis?

Answer: The first step is to remember that bulk density can also be represented as g/cm^3, thus for this soil the bulk densities for this unit are 1.2 and 1.5 g/cm^3 (which is the same as g/cc.) The next step is to decide how many grams of soil are needed in the sample. If each sample is to contain

50 g the equation to make the required calculation is:

Bulk density × area × depth = 50 g

rearranging the equation

$$\text{Depth} = \frac{50 \text{ g}}{\text{Bulk density} \times \text{area}}$$

Carrying out the calculation for the soil with a bulk density of 1.2 g/cc the depth of the sample taken would be 8.3 cm and the sample with bulk density of 1.5 g/cc would be 6.6 cm.

The bulk density of soil horizons and underlying layers is important in determining the volume of soil needed during the sampling process. In addition, knowing the bulk density of the horizons can indicate the likely path of water and contamination through the soil (see Figure 2.8) [19].

2.5.　WATER MOVEMENT OVER AND THROUGH SOIL

Water is the great mover in the environment. Air and ice also move soil; however, air cannot directly move particles larger than sand, while water and ice can. Although ice can move large boulders, it is not a major component of most of the environment. Water and ice move very large rocks and boulders, and in this process they are reduced to gravel, sand, and silt, and clay may be released if present in the rock. Ground rock is deposited on banks along rivers and streams and at the mouth of rivers and the end of glaciers.

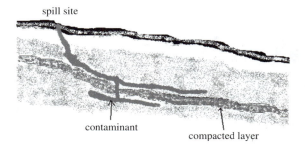

FIGURE 2.8 Water moving downward and horizontally in soil containing compacted layers.

Rivers and streams may bury materials, including contaminants. When a river confined in a valley comes to an open plane its velocity will decrease and the material it is carrying will be deposited. This gives rise to

FIGURE 2.9 A soil profile, knife at the bottom of the A horizon. The very dark material at the bottom (below white line) is buried material.

what are called alluvial fans, which are similar to deltas except that they are deposited on dry land. In addition to simply carrying material and depositing it, streams and rivers tend to meander. Over a number of years a stream might move from one side of its valley to the other, completely reworking the landscape in the process. Contaminated materials may be buried in all three of these situations. Figure 2.9 shows a soil profile with buried material.

In practical terms this means that contaminated material in the stream may be buried anyplace in the floodplain (Figure 2.10). A field sampling plan must take this potential burying of contamination into consideration. Also, all floodplains are subject to flooding, and thus are not appropriate places to store hazardous materials or to set up field office buildings, particularly those for storage of equipment and samples.

Also as seen in Figure 2.10, a contaminant on a hillside may be closer to the soil surface at the top of the hill and deeper lower on the hill, because erosion will remove soil from the top of the hill and deposit it at the bottom. Gravity can also move soil down a slope and thus tends to bury contaminated areas. Soil moved under the influence of gravity is said to be colluvial. Figure 2.11 shows a hillside where material has moved under the influence of gravity.

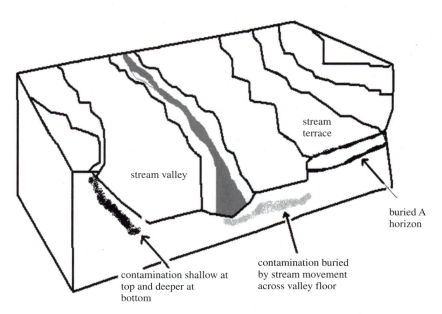

FIGURE 2.10 Stream terraces and stream meandering across a stream valley.

FIGURE 2.11 Soil moving down a bank under the influence of gravity (colluvial). Note the tree (A), which has moved with the soil.

In addition to moving over the surface of the soil, water percolates down through the soil into the underlying loose material and rock. Salts, amendments, and spills will tend to move down through the soil and underlying material with the water. Horizons in soil, their texture, structure, and density, will be important in determining how the water and material it is carrying moves. If at any place in the soil profile or in underlying material there is a change, water will stop and move laterally along the boundary. (See Figure 2.8.)

In Figure 2.12 there is a sand layer overlain by a gravel layer (gravel does not pass a #10 sieve), which in turn is overlain by a sand layer. Water moves into the upper sand layer and fills this upper layer before it moves through the gravel layer. Sand has finer pores or capillaries than the gravel. Water can move from large into smaller pores or capillaries, but without an

FIGURE 2.12 Movement of water through a soil having layers of differing texture and composition.

energy source it cannot move from small to large. Thus, the water will not move into the gravel layer unless there is enough pressure exerted by water buildup to move it from the small sand pores into the larger gravel pores.

Lateral movement along boundaries, as shown in Figure 2.8, will continue until one of two things happens: the water encounters a change in the horizon that will allow deeper movement or the water above the horizon accumulates to such an extent that it exerts enough pressure to push the water lower. In Figure 2.8 water-carrying contaminants encounter a compacted layer that prevents deeper movement. When a hole in the compacted layer is encountered the water moves deeper. Note that it still continues to move further along the compacted layer. Also, once through the compacted layer it moves in both directions (left and right). Almost any change in the composition or characteristics between layers in the soil or underlying material will cause water to move laterally.

Another important factor in the movement of water in soil is plants. Soil having plants growing on it will have higher infiltration rates than bare soil. Even coarse sandy or gravely soils will have higher infiltration rates when plants are growing on them.

2.6. CONCLUSIONS

In field sampling it is important to keep all aspects of the environment in mind. The atmosphere can have significant effects on sampling as a result of windblown contaminants, photochemical reactions, and the exposure of anaerobic samples to oxygen. In terms of the hydrosphere, water is the great mover in the environment and is important in predicting where contaminants will be found. The lithosphere will either affect or control the movement of water, particularly through the soil characteristics of texture, structure, and bulk density, and the rock characteristics of porosity and cracks. Thus, it is important to know the characteristics of soil and rock in order to know where and how to sample. All three parts of the environment are intimately interconnected and thus are important in field sampling. (The characteristics of the environment described in this chapter will be necessary inputs for the environmental models described in Chapter 7.)

QUESTIONS

1. List the three major spheres of the Earth.
2. Name two surface and two subsurface horizons commonly found in soil. Give some important characteristics of each horizon.
3. Using the textural triangle find the percentage of clay in a soil sample having 85% sand and 3% silt. What should the textural name of such a soil be?
4. What would the engineering letter designation for the soil in question 3 probably be?
5. Two soil samples—ARC32a and ARC32b—are taken from a field while traversing it. One has a bulk density of 0.95, and the other a bulk density of $1.25\,\text{Mg/m}^3$. What is the major constituent of these two samples?
6. From just the bulk densities can you tell what happened between the two samples?
7. A field sample is given an engineering designation of SM. What does this correspond to in the textural triangle?
8. Give two examples of how one of the Earth's spheres might affect another in a way that would impact sampling activities.

9. Using diagrams and words describe the various layers of loose material on the Earth's surface.
10. In the diagram in question 9 where would the capillary fringe be located?
11. A soil bulk density sample is taken. The ring containing the soil has a diameter of 6 cm and a height of 3 cm. The soil plus ring weight is 114.9 g, and the ring weight is 15 g. Determine the bulk density of this soil and its percentage solid portion and percentage void space.
12. Describe the relationship between atmospheric air and soil air.
13. What is the effect of plants on the infiltration of water into soil?

REFERENCES

1. Gbonbo-Tugbawa SS, Driscoll CJ. Retrospective analysis of the response of soil and stream chemistry of a northern forest ecosystem to atmospheric emission controls from the 1970 and 1990 amendments of the Clean Air Act. Environ Sci Tech 2002; 36:4714–4720.
2. Martin W, Nunnally NR. Air and Water: An Introduction to the Atmosphere and Hydrosphere. Dubuque, IA: Kendall/Hunt, 1999.
3. Pierzynski GM, Sims JT, Vance GF. Soils and Environmental Quality. 2nd ed. New York: CRC, 2000:22–26.
4. Schwab GO, Fangmeier DD, Elliot WJ. Soil and Water Management Systems. 4th ed. New York: Wiley, 1996:144.
5. Schrader W, Geiger J, Klockow D, Korte EH. Degradation of α-Pinene on Tenax during sample storage: Effects of daylight radiation and temperature. Environ Sci Tech 2001; 35:2717–2720.
6. Johnson CA, Leinz RW, Grimes DJ, Rye RO. Photochemical changes in cyanide speciation in drainage from a precious metal ore heap. Environ Sci Tech 2002; 36:840–845.
7. Brady NC, Weil RR. The Nature and Properties of Soils. 12th ed. Upper Saddle River, NJ: Prentice-Hall, 1999:485–521.
8. Rendig VV, Taylor HM. Principles of Soil–Plant Interrelationship. New York: McGraw-Hill, 1989:53–57.
9. Hromnik CA. Karst, Kas er Karasattu: Whence the name? Cave and Karst Science. London: British Cave Research Association, 2001; 28:79–88.
10. Jury WA, Gardner WR, Gardner WH. Soil Physics. 5th ed. New York: Wiley, 1991:196.
11. Pierzynski GM, Sims JT, Vance GF. Soils and Environmental Quality. 2nd ed. New York: CRC, 2000.
12. Strahler AN, Strahler AH. Introduction to Environmental Science. New York: Wiley, 1974:68.
13. Kling GW. Seasonal mixing and catastrophic degassing in tropical lakes Cameroon, West Africa. Science 1987; 237:1022–1024.

14. Massei N, Lacroix M, Wang HQ, Mahler BJ, Dupont JP. Transport of suspended soils from a karstic to an alluvial aquifer: The role of the karst/alluvium interface. J Hydrol 2002; 260:88–101.

15. Soil Survey Staff. Soil Taxonomy: A Basic System of Soil Classification for Making and Interpreting Soil Surveys. Soil Conservation Service, U.S. Department of Agriculture. Agriculture handbook no. 436. Washington, DC: U.S. Government Printing Office, 1975.

16. Buol SW, Hole FD, McCracken RJ. Soil Genesis and Classification. Ames, IA: Iowa State University Press, 1980.

17. Foth HD. Fundamentals of Soil Science. 8th ed. New York: Wiley, 1990:24–26.

18. Soils Manual for Design of Asphalt Pavement Structures. manual series no. 10 (MS-10). College Park, MD: Asphalt Institute, 1969:69–78.

19. Brady NC, Weil RR. The Nature and Properties of Soils. 12th ed. Upper Saddle River, NJ: Prentice Hall, 1999:125–149.

3

Presampling

Before a physical sample is actually taken the whole sampling procedure and all required equipment must be in hand and ready to use. For large projects a field office, laboratory, and storage area must be in place and functioning at the field. The sampler or samplers to be used must be at the field, along with the project notebook, the sampling plan or diagram, maps of the area, and suitable containers for samples. The sample storage and transport containers must be present and ready to use, along with chain of custody and other forms. Plans for shipment and analysis must have already been discussed and planned with the commercial laboratory doing the analysis. (See Chapter 10 for a more complete description of this process.)

A minimal field office will include not only a building with two main rooms (office and laboratory) and a washroom; a sample storage area separate from any of these areas must also be available. Both the office and laboratory will need to be equipped (chairs, desks, computers, etc. for the office, and laboratory benches and standard laboratory equipment—e.g., balances, mixers, pH meters, and other meters or measuring devices needed for the particular sampling situation). These will all need to be installed and functioning before sampling begins.

Other tools to be used during the sampling will also need to be acquired. For instance, if ground-penetrating radar (GPR) is to be used, the equipment and operators must be contracted. Suitable GPS units, computer programs for statistical analysis, and geographic information display (GIS) will also need to be purchased. Personnel will need to be educated or trained in both the use of these tools and the interpretation of the data produced. If inexperienced personnel will be sampling, they need some specific training in proper sampling techniques. (See Chapter 11 for further discussion of personnel.)

3.1. FIELD OFFICE

A field office consisting of one or several buildings with electricity, phone service, and running water will be set up at the field. This building is usually a trailer with one to three rooms. One room will be for such things as the computer, phone, project notebook, and chain of custody and other forms. Two phone lines are recommended because one will be used for the modem in the computer. A second room will be set up as a small laboratory, in which needed field analytical procedures can be carried out. This will be particularly useful during transect sampling and for testing and analysis that needs to be accomplished immediately. A bathroom for cleanup is essential, along with an outside shower for situations in which sampling personnel become massively contaminated. It will also be useful for washing off boots and equipment after sampling.

A separate building is needed for storing samples and another is needed for storing equipment, including any needed chemicals. Note that chemical storage must be well separated from sample storage to prevent cross-contamination. Any storage building designated for chemicals needs to have ventilation and an explosion-proof fan for moving air through and removing fumes.

In a rural or farming-type setting buildings may already be at the field and available for use. Such buildings may already have electricity, water, and bathroom facilities and be suitable for use as the office and field laboratory. However, it is preferable to have a separate sample storage building. Farm buildings are used to store all manner of chemicals, from fertilizers to pesticides to antibiotics to disinfecting chemicals. Fumes from any of these sources can be absorbed by field samples, contaminating them and causing complications in analytical procedures later. This could result in the analytical results showing contamination that is not in the field but is actually in the storage building. In a worst-case situation a recomendation may be made for a field to be remediated for a contaminant that it does not contain.

The importance of a separate room that has never been used before for sample storage cannot be overemphasized. Even if a room or building has been thoroughly cleaned and has no smell or other evidence of contamination it can still be there. Colorless and odorless contamination can be contained in cracks and dust in the walls, floor, and ceiling and can be just as detrimental to samples as any other contamination.

3.2. FIELD OFFICE LABORATORY

Many field sampling situations will call for setting up a small laboratory at the field office. This should not be a table in a corner of a busy area or where samples for shipment to commercial analytical laboratories are stored. This is an area that needs to be in its own room with running water, electricity, an explosion-proof exhaust fan, and sufficient bench space to allow for setting up instruments and working with samples.

The laboratory will be provided with a minimal amount of common laboratory equipment. Any work done in any laboratory will require purified water, which can either be distilled in the laboratory or purchased. Another source is water that has gone though several ion exchangers and reverse osmosis processes and is called deionized (DI) water. Distillation equipment is usually bought, while DI/reverse osmosis equipment can be rented. This water will be used for preparing solutions and for final cleaning of glassware.

Soil samples will require drying and sieving. Drying is usually done by placing a thin layer of the soil sample on a clean surface, such as a piece of plastic, and allowing it to air dry. It will always be more convenient to dry the sample using an oven. Unfortunately, drying a soil sample at even slightly elevated temperatures causes irreversible changes, thus soil samples should never be mechanically dried before analysis.

In most cases a #10 and #200 sieve will be sufficient for most sieving needs. It is good to have a receiving can or pan to place under the sieve to catch sieved material. Typically sieves are constructed of brass with soldered-in mesh. This means that sieved samples can pick up metals such as copper, zinc, lead, and silver during sieving. Stainless steel and plastic sieves are also available for more demanding situations in which metal contamination must be avoided. Metals will normally not be of concern for the samples analyzed at the field laboratory. In cases in which samples are to be sent to a commercial laboratory for further analysis, however, contact with metals will be of concern.

Other equipment needed will include a balance and dissolved oxygen and pH meters. At minimum a top-loading electronic balance having a capacity of several kilograms and a readability of at least 0.01 g will be needed. Balances with readability of 0.001 g are readily available, but such sensitivity is not generally needed. A pH meter will be needed in all field sampling situations, but a dissolved oxygen meter will only be required when sampling water or water-saturated samples. Inexpensive portable and handheld pH meters are readily available. The more versatile multi-functional pH meters are more useful and usually more durable, however. These meters can measure pH and temperature and accommodate ion selective electrodes (ISE). Ion selective electrodes are available for measuring ammonia, a wide range of metal cations, and anions, including cyanide, nitrate, nitrite, and many others. These types of electrodes and measurements can be very useful, particularly during the transect sampling phase.

Using pH and other electrodes requires filling and standardizing solutions. Most of the required solutions can be obtained premade, and this is recommended. In no case should electrode filling solutions be made at the field laboratory. Standardizing solutions for pH and ISE can be prepared at the field laboratory from premeasured packets of the required solutes. These are dissolved in the prescribed amount of distilled or DI water, which should be freshly prepared for this purpose to assure high quality.

An assortment of typical laboratory equipment will also be needed. A shaker for shaking samples, an oven, and a rack for drying glassware will be useful in the laboratory. Also, Erlenmeyer flasks, beakers, graduated cylinders, spatulas, weighing paper, wash bottles, brushes, and laboratory cleaning detergents must be available. In some cases, other project-specific equipment (e.g., flammable gas detectors for areas contaminated with fuels) will be needed. Some forethought as to the equipment needed will allow smooth operation of the field office laboratory. A summary of general equipment and supplies needed for a field laboratory are given in Table 3.1.

Unless there is some important reason for doing otherwise, only minimal analysis should be undertaken at the field laboratory. Instrumentation, conditions, and knowledge are needed for proper operation of sophisticated instrumentation such as spectrometers and gas chromatographs. Analysis requiring these instruments should be left to the commercial analytical laboratory.

The operation of any laboratory entails the generation of waste. The waste may be in the form of broken glassware, discarded excess sample material, chemicals used in analysis, and innocuous materials, such as paper towels for drying hands. Broken glassware is the source of cuts in the laboratory, and if not disposed of properly, may be hazardous to persons handling trash. Special broken-glass containers are available and should be used for all broken glass. Any samples containing hazardous material either from samples or from analytical procedures should be disposed of properly. Other material can be disposed of in local landfills.

TABLE 3.1 Field Laboratory Equipment and Supplies

Equipment/supply	Comments
Room	Separated from other work areas and containing bench space, running water, electricity, and explosion-proof exhaust fan
Purified water source	Distilled or deionized water
Bench space	For air drying samples
Sieves	#10 and #200; others may also be needed
Balance	Capacity kilograms, readability 0.01 g
pH meter	Capable of ISE measurement
Dissolved oxygen (DO) meter	Portable meter preferable
ISE electrodes	For the contaminants expected
Standardizing solutions	For pH, DO, ISE
Shaker	Reciprocal to hold Erlenmeyer flask
Drying oven	Medium size for glassware
Other as needed	Beakers, flask, spatulas, graduated cylinders, weighing paper, bottles, etc.
Waste disposal containers	Must be capable of safe disposal of broken glass

More information about laboratory procedures and analytical methods as they relate to field sampling are given in Chapter 10.

3.3. THE PROJECT NOTEBOOK

The project notebook (sometimes called a field notebook) is a book in which all information about the project is placed. It is intended for use in the field and is the place in which all notes, diagrams, sketches, maps, and information about the project are kept. This is traditionally a bound book into which additional notes, maps, and other information are inserted and fixed. The fixing should be done with either glue or tape, not paper clips or staples.

The most important problem with the project book is neatness. It is good to have a project book be as neat as possible. In too many instances, however, workers fail to put important data in the project book because they are afraid that it will not be neat. They would rather write notes on scraps of paper and then write a neat entry in the project notebook. Sometimes people wish to write in pencil so that mistakes can be erased. It is important both for completeness and legal reasons that none of the above be allowed.

All data, observations, and any other pertinent information must be put directly and permanently into the project notebook with a nonsmearing pen. Although neatness is preferred, it is not unprofessional for smudges and crossed out (but not "whited" out) entries to occur on some or even all pages. For both practical and legal purposes it is better for the project notebook to be a little less neat but absolutely complete. In Figure 3.1 the entry on line 7 has a part marked out. Entries or even whole pages can be crossed out, but pages are never taken out of the project notebook. Missing pages are an indication that serious problems exist. The project notebook should thus be bound rather than loose-leaf, which means that for large projects there may need to be several volumes of the project notebook.

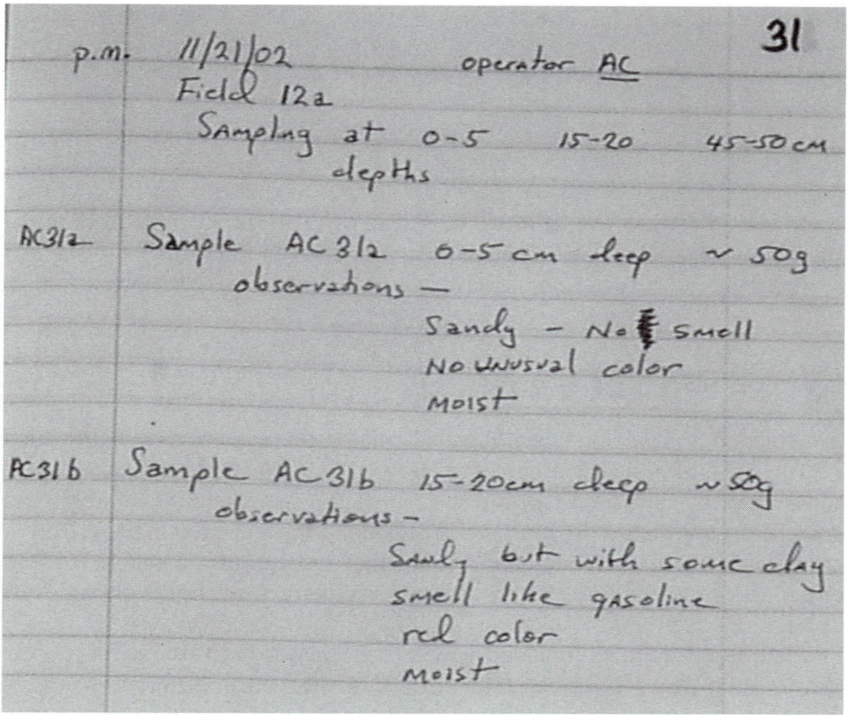
31

p.m. 11/21/02 operator AC
Field 12a
Sampling at 0-5 15-20 45-50 cm
depths

AC31a Sample AC 31a 0-5 cm deep ~ 50g
observations —
Sandy — No smell
No unusual color
Moist

AC31b Sample AC 31b 15-20cm deep ~ 50g
observations —
Sandy but with some clay
smell like gasoline
red color
Moist

FIGURE 3.1 An example of a project notebook.

A standard way of handling entries is to first number all the pages. In Figure 3.1 the page number is in the upper right-hand corner. Second, it is highly recommended that five to ten pages at the beginning of the book be left blank. These pages will be used as a table of contents. As work progresses, samples are taken, observations are made, and results are obtained, these will be written up in the body of the book and noted in the table of contents. The table of contents tells where these data are located so that when a report is needed later the required information is easy to find. The table of contents is also handy in connecting observations and data collected at different times and bringing them together. Samples will be taken, and two weeks later the analytical results will be returned. In the meantime entries have been made in the project book. Information about how and when the samples were taken will need to be combined with the analytical results in reports, so the ability to readily access the sampling data and the analytical results is important.

The date on which entries are made in the project notebook must be placed at the top of each page. In some cases it may also be necessary to note the times at which activities were undertaken either at the top of the page or in the body of the write-up. In Figure 3.1 the date is at the top of the page on the left and follows a notation that the samples were taken in the afternoon (i.e., p.m.).

When samples are taken, the numbers they are given should refer back to the project book. Specifically, the sample numbers should refer back to the page number where a description of the sampling process and the location of the sample are given. It is usually a good idea to prefix the sample number with the initials of the person taking the samples. In addition, people often like to have the sample number include a reference to date and time. For instance, the sample number might be AC12A31a112102pm. Here 12A is the field number, 31a the sample number in the project book, 112102 the date, and pm the time of sampling. As can be seen, this can produce an unwieldy sample number if carried too far.

In Figure 3.1 the sampler's initials are at the top of the page and each sample entry begins with his initials followed by the page number. The first entry has a small letter *a* following the sample number. This tells the reader that it is the first sample noted on this page. Also note that the sample number is repeated in the left-hand margin for easy reference. Under each sample number various observations are made about the sample as it is obtained. Such observations can add critical insight into the nature of the sample and its components.

An example of a simple observation is the color of a soil sample, which can be important in interpreting sampling and analytical results. A soil sample showing color variations of reds, browns, grays, or blues indicates that the soil is under reducing conditions during a significant period of time during the year. The ease or difficulty of inserting the sampling tool is another observation that can indicate the occurrence of compacted layers. Samples might also show a change in texture or the occurrence of a gravel or sandy layer. As mentioned in Chapter 2, all of these are important to note because they can indicate where additional sampling may be needed or where a contaminant is likely to be found.

In a similar fashion observations about water and air samples must be made as they are taken. These may include color, suspended matter, and odor, and may help in identification of the nature and concentration of the contaminant or contaminants. Observations will also provide either clues or evidence as to the source of the contamination, which could be essential in preventing further contamination.

3.3.1. Number Changes

Sometimes another sample number replaces the number given in the field when the sample is processed for storage, shipping, or almost always, in the commercial laboratory. Renumbering opens up the whole process to error and is not recommended. If, however, it cannot be avoided it is strongly recommended that the original sample number be kept on or with the sample even after a new number is given. In all these cases a cross-reference list relating the two different numbers must be readily available and a copy must be kept with the samples at all times.

Sample numbering and recording process mistakes will be made during the sampling. Numbers will not be written legibly, or they will get erased or smeared and become unreadable. To the extent possible every effort should be made to eliminate or limit these types of errors. Using nonsmearing pens, using paper that can be written on when wet, and putting the sample numbers on the sample containers in several places are all good approaches to limiting these problems. It is far less expensive and less time consuming to spend extra time on sample labeling than on taking another sample or assuming a sample's identity [1].

One way to help minimize sample labeling errors is bar coding. There are software programs, computers, and printers that can print bar codes on various media, including labels of different sizes. Not only will bar codes help prevent mistakes in labeling, they will also speed up sample handling, especially in maintaining the chain of custody. If bar codes are used, a bar code for the sample needs to be in the project notebook in which the sample is described (see Chapter 8 for more information on bar coding) [2].

A GPS unit will often be used in sampling a field. This will give the location of a sample site and may even record a sample number to be associated with that site. Even in this case it is important to have a project notebook to record these and other data about the sample when the sample is taken. At the time of sampling this will seem to be unnecessary repetition, while a week later it will be found to be an essential, time-saving step.

3.4. MAPS

Maps are the heart of the sampling process and are essential for any successful sampling program. There are many types of maps available; for example, surface, topographic, and three-dimensional. Especially valuable are aerial photographs of the specific area to be sampled. If they are not available there are both firms and computer programs that can be called upon to produce the needed maps, aerial photographs, or both. Before doing anything at a contaminated field it is essential to have several maps of different scales of the area. One map should show the contaminated field and the surrounding areas. Several maps showing details of the field itself will also be needed.

I would not, however, recommend producing a large number of very detailed and expensive maps. What is needed are many simple maps of the area suitable for recording data on. Each time some activity takes place on the field it needs to be recorded not only in the project notebook but also on a field map, so that the location is precisely known. The place in which the activity takes place must be on the map, along with date and time and the name of the person leading the activity. It can be separate from the project notebook during the activity, but either the original or a copy must be permanently attached to the project notebook in the appropriate location when the activity is completed. Each map should have its own entry in the table of contents. Alternately, a separate page listing just the location of maps can be in the table of contents. Data recorded on a map can subsequently be entered into a computer either manually or by scanning the information. Once in a digital form GIS can be used to display and follow sampling and other field activity.

An example of a simple map of a field to be sampled is shown in Figure 3.2, and includes a symbol showing which direction is north (which must always be present). The map must also show some distinct characteristic that allows the reader to specifically locate it. In Figure 3.2 the roads specifically locate the position of the fields of interest, provided, of course, that we know the state in which these roads are located. Additional locators such as longitude and latitude can be included on the map and are recommended. These are useful especially when using GPS for locating the

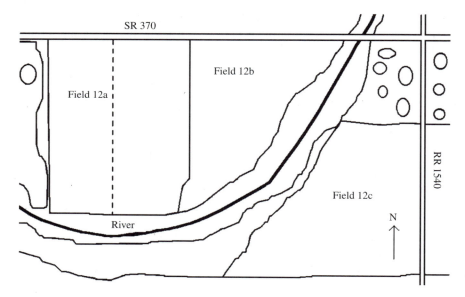

FIGURE 3.2   Simple map of three fields (12a, 12b, and 12c) to be sampled.

fields and sample sites and can be entered from the **GPS** unit during sampling.

A cross-sectional map of soil depths and the depths of various layers (see Figure 3.3) is also valuable to have. Such a map is helpful in developing

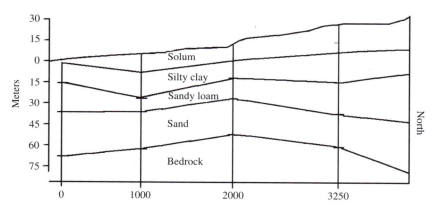

FIGURE 3.3   Cross-sectional map of the profile of regolith in field 12a. Units at the bottom of the map are meters.

both the transect and detailed sampling plan. It can help in determining where a contaminant may have moved and thus where more samples need to be taken. Such a map may be developed from data in the soil survey of the field, which is discussed below.

Another useful type of map is the topographical or topo map of the area. An example of a topo map is given in Figure 3.4. The topo map has lines of equal altitude drawn on it. They thus show slopes, escarpments, and other important features of a landscape. They also show which direction water will flow in and where erosion is most likely. In many cases topological characteristics will need to be taken into consideration when sampling.

For many areas the United States Geological Survey and other government entities have digital orthophoto quarter quadrangles (DOQQ) photographs or maps of areas of interest. These are digital photographs that are placed on the Universal Transverse Mercator Projection (UTM) on the North American Datum (NAD-27). There is also a NAD-83, and others, that can be used. Digital orthophoto quarter quadrangles have a ground resolution of 1 meter and cover an area of 3.75 minutes latitude by 3.75 minutes longitude. It is useful to note that both UTM and NAD-27 are used by GPS units, and since they are digital, they are compatible with

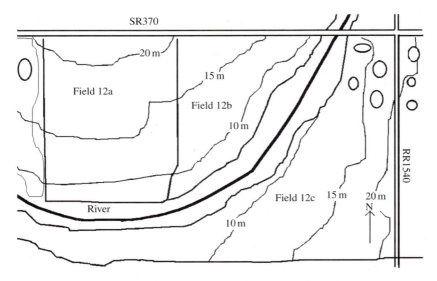

FIGURE 3.4 A contour map of fields 12a, 12b, and 12c showing contour lines which give the elevation of different parts of the field.

FIGURE 3.5   An example of a portion of a DOQQ map from Clinton County, Ohio, showing roads, buildings, and fields.

GIS systems An example of a portion of a DOQQ map is shown in Figure 3.5 [3,4].

The DOQQ images produced have both the look of a photograph and the geometric qualities of a map. They are a true representation of the Earth's surface. Because of this both distances and areas can be determined directly from the DOQQ. Software for manipulating these maps (e.g., MrSID GeoViewer) is available—some of it free—on the Internet [5].

Digital orthophoto quarter quadrangles are very useful in sampling, and are available at a minimal cost. Unfortunately they are often hard to find, even on the Internet. One relatively easy place to find information about DOQQs is the local GIS office. Many county courthouses will have a GIS office, and personnel doing GIS for the county will be very helpful in obtaining needed DOQQs. If no county GIS office is available, try the state GIS office or officer.

When sampling, site positions can be determined using a GPS unit. These site positions are saved and later loaded into a GIS system. A DOQQ is then also loaded into the GIS system. The system is then used to generate a map that shows the true location of sample sites in the field. This map and a GPS unit can be used to locate the sampling sites at any time in the future.

It is also possible to enter other sampling data, such as sample depth, altitude, vegetation cover, and other information deemed necessary into the GIS system. These data can also be displayed on the thematic map, and all the data can be interrelated as needed or desired.

3.4.1. Soil Survey

The soil survey is another invaluable map to have before sampling a field begins. This is a map of soils in an area, but it also contains additional information about the area it represents. The soil survey is made up of aerial photographs of an area of the Earth's surface with the designations of the soils in the field indicated on it, as shown in Figure 1.2 in Chapter 1. Descriptions of the soils, their characteristics, and their suitability to various uses are also given, as are characteristics of the landscape, such as slope, buildings, waterways, and roads. This is an extremely useful type of map and source of information and no field sampling project should be undertaken without having a soil survey of the area available for ready reference. In situations in which the field sampling is primarily of water or air a soil survey and associated maps are still useful.

Soil surveys have typically been carried out only in rural areas; that is, they have been seen as tools for agriculture, particularly in erosion control. In areas of rapid urbanization older soils maps may show the soils in an area because it was agricultural land when the map was made. New soils maps attempt to include all land in the soil survey, thus both the older and newer soils maps can be valuable in a field sampling project.

Field sampling will include sampling a similar uncontaminated field or soil. The soil survey will be indispensable in finding uncontaminated soils, which are the same as those in the field under investigation. Such comparison soil samples are essential to be certain that the analytical results represent an unusual situation in the field of concern. It is often important to know the base or natural level of a component so that it is not mistaken for a contaminated level. See Chapter 9 for an introduction to the types of elements and inorganic and organic compounds likely to be found in soil.

3.5. PRELIMINARY FIELD SURVEY

Before any sampling an initial survey of the field should be carried out. This means looking at the field from all directions. The first step is to walk completely around the field looking at it from all aspects and noting all the conditions in and around it. During this process keep asking what is seen and what is going on. Another important question to keep in mind is whether there are additional sources of contamination that might affect the sampling, analytical results, or the cleanup plan.

Field characteristics that should be noted during this first survey are given in Table 3.2. The topics in this list and others, such as the clay and clay type, bulk density, roots, pH, electrical conductivity, and organic matter content, will be given in the local soil survey. In spite of this it is a good idea to check these data by making observations in the field, because significant changes may have occurred between the time the soil survey was made and the sampling undertaken.

An example of such a situation occurred at a Superfund field. This field was 1 ha in size. On the upslope side of the area was a railroad siding consisting of three sets of tracks that carried heavy freight train traffic. On another side was an electrical power substation. Of concern here would be heavy metals, grease, tar, and creosote from the railroad. This was of particular concern because the tracks were upslope from the site. The other concern would be polychlorinated biphenyl (PCB) contamination from the electrical substation.

Although this is a relatively simple and obvious example it is not an unusual situation. In this case routine sampling and analysis might not show additional needed cleanup and remediation work. On the other hand, a routine analysis might have indicated additional contamination requiring additional sampling and analysis, which could have been minimized by recognizing the sources of additional and continuing contamination [6]. Another similar problem is that different, unrelated contaminants may interfere with each other in analytical procedures. This could mean that the analytical laboratory might obtain confusing results related to the additional sources of contamination from a rail line or a power station. (See Chapters 10 and 11.) Knowing what is around the field and other

TABLE 3.2 Types of Field Characteristics to Be Noted During Preliminary Survey

Characteristic	Types of observation
Slope	Steep, gentle, flat, depressional[a]
Climate	Arid, semihumid, humid
Wind direction	Note predominant direction
Vegetation	Luxurious, sparse, crops, grass, trees
Erosion	Obvious evidence of erosion? Rills, gullies, etc.
Water	Evidence of water movement into, over, or through area
Contamination	Sources of contamination and continuing contamination

[a]An area or field in which water from surrounding areas collects during a rain event. Often this area does not have an outlet.

possible sources of contamination will make discovering and understanding these sources of confusion much easier.

As with all activities related to field sampling notes about this initial survey must be put in the project notebook. All observations about the field need to be entered with particular reference as to where the items of note are located in the field, and as indicated above these observations are to be included on the appropriate maps of the field and their positions marked using GPS.

3.6. FIELD HISTORY

After surveying the field the second thing to be put in the project book is a field history. The best field history is one that includes all available information about the field use or uses before its latest use and the use that has necessitated the sampling or cleanup. The sources of this information will be owner records, along with deeds of sale and municipal, county, state, and governmental records and any available GIS thematic maps. In some cases it may also be valuable to talk with people living in and around the field about what has happened there. Do not assume that an isolated field on a back road in the country is not contaminated. I have seen hidden fields 20 miles from a city being used clandestinely as dumpsites. Also, isolated areas on a farm might be used to dispose of agricultural chemicals.

As indicated above, it is also a good idea to know what the areas around the site were being used for historically. These surrounding areas can be a source of additional contamination. (e.g., low-lying areas can become contaminated as water washes contaminants or eroding soil from surrounding or neighboring fields into the area).

Field History in Relationship to Phosphate Content

I often work on projects in developing countries. One project involved a search for a source of phosphate, which might have been used as fertilizer. I was supplied five soil samples from an area, composed of several fields, which were suspected to contain a phosphate rock deposit. Analysis of the samples gave mixed results. Some samples had undetectable amounts of phosphate, and other samples had moderate to high levels of phosphate.

In this case a lot of money would have been saved if a detailed history of the fields had been obtained. A map showing the location of previous activities in the fields and the sampling sites before detailed sampling and analysis was done would also have been useful. It turned out that the samples were taken from fields that had at one time been a village and at another time a burial site. Soil samples that came from cooking fire sites showed moderate levels of phosphate. Samples containing chips of bone showed high levels of phosphate. Soil not contaminated with ash or bone had undetectable levels of phosphate.

Often historical information about a site is not as complete as one might wish. It is not unreasonable to talk with people who have worked on various projects that were carried out at the field and with people living around the area. Sometimes some surprising activities can come to light during this investigative phase.

The purpose of this phase of investigation is to understand the sampling needs, provide proper safety for workers, and avoid unpleasant surprises. The potential depth and extent of sampling will be indicated during this phase, as will the type of sampler or samplers needed. The types of samplers used will depend on the composition of the field, the type of component to be determined, and the depth of sampling. In a field requiring small numbers of shallow samples a hand sampler capable of sampling 3 meters deep may be sufficient. If more or deeper sampling is required, powered samplers or drills may be needed [6].

Without knowing the history of the field it is hard to anticipate such safety needs as what or how much protective and emergency equipment is needed. Will safety glasses and work boots be sufficient or will more sophisticated coveralls, rubber gloves, respirators, and so on be needed? There is also an economic aspect to knowing the history of the field in that the amount of safety equipment needed will change, depending on the level and the type of contamination. If there is no contamination and the sampling is for other purposes, minimal safety protective and emergency equipment will be needed [7,8].

Another aspect of obtaining the history of the field is to avoid unpleasant surprises. One such surprise might be discovering underground tanks containing unknown waste. Another might be discovering unsuspected contaminants that affect the sampling and remediation plans. It is far less expensive to know about these beforehand than to have to contend with them later.

One last piece of information to obtain is to identify all possible sources of contamination or pollution. Sources may be point sources, so that the exact location of the source of the contamination, such as a spill of fuel oil from a tank car, can be pinpointed. The other is the nonpoint source. In this case the contaminant may be spread out over a large area, such as an agricultural field of several hundred acres. This diffuse material may wash off the soil surface and be concentrated in low areas or bodies of water. It is much easier to find, identify, sample, and remediate point sources of contamination than nonpoint sources, but it is essential to identify all possible sources as early as possible.

Surprises in the Field

At one cleanup field that I was assured had been previously investigated, I found an unpleasant surprise, which is the one thing I do not wish to find at a field. I attempted to take a soil sample only to encounter a board 5 cm under the surface. Five cm deeper I found a brick, and under that a large rock. I was immediately concerned. What had been dumped in this area? If it was fill, the question was, from where? Was the fill clean to begin with? Had this been a dump at one time? It turned out that relatively clean fill had been used. This is one example of surprises that can be unearthed if an adequate history of the field is not obtained.

As part of the history of the field find out about the soils present and their characteristics. This information can be obtained from soil surveys. If the soil is not the same as indicated in the soil survey, it is either fill or disturbed by various activities on the field. Knowing the soil—particularly its texture—will help determine the type of sampler needed. This information will also be required to determine how deep sampling must be done or is likely to be needed. The soil survey will also indicate if water is likely to be encountered during sampling [9]. Information in the soil survey will also be needed in any modeling of the field.

3.7. SAMPLING TOOLS

During the presampling stage decisions about sampling tools and their sources need to be made. Solid materials, particularly soils, will require different samplers, depending on their characteristics. Water and air will each require different samplers, although in many cases no sampling tool will be needed.

On the basis of the characteristics of the soils present or likely to be present as indicated by the soil survey, what type of sampler is likely to be most useful? In part this will be determined not only by the texture of the soil but also the moisture levels likely to be encountered. Dry, sandy soils will require a different type of sampler than wet, clayey soils. Likewise, a stony or rocky soil will require yet another type of sampler. Another important consideration is the depth to which sampling is to occur. The sampler must be able to sample to the maximum depths needed to account for all the contaminant likely to be present. (See below and Chapter 5 for more detailed information about samplers.)

Contaminant type, whether inorganic or organic, will affect the choice of sampler. For inorganic components the metal used in the sampler will be important. If, for instance, zinc is the component of interest the sampler must not have zinc in it. Many different types of metals are used in the construction of samplers, so it is important to be sure that the composition of the sampler is compatible with the sample being taken. If the component is organic, there are two other considerations. The first is that the sampler (or sampler liner, if used) is not adding organic material to the sample. The second occurs when the sample contains volatile organic compounds (VOCs). In this case the sampler should, as much as possible, minimize the loss of the organic components during the sampling process [9].

3.7.1. Soil Samplers

There are three different basic types of soil sampler heads—simple auger, bucket auger, and core. The three different types are shown in Figure 3.6. (See also Figures 5.5 and 5.6 in Chapter 5.) A simple auger looks like an old-fashioned wood drill bit. It can be used in moist and wet fine-textured and rocky soils but does not work well with sandy soils, particularly when they are dry. Auger samplers will penetrate rocky soils and allow sampling of finer fractions, but will not remove larger stones. With extension handles and proper soil conditions it can be used to sample up to 6 meters deep. If many deep samples are to be taken, however, using this sampler will require a lot of work and be very time-consuming.

The bucket-type auger is effective for sandy, muddy, and rocky soils. Hand-operated samplers of this type can sample up to 6 meters deep, but need extension handles and the proper soil conditions, along with a lot of effort and time for sampling. Rocks will sometimes cause problems if they get caught in the blades of the sampler. In such a case they prevent further sampling, and their removal can interfere with the sampling process and obtaining an adequate sample. Larger rocks can prevent this sampler from penetrating the soil and thus prevent a sample from being obtained.

There are a number of types of core samplers, which consist of a hollow tube inserted into the soil. When the tube is withdrawn it contains a core of soil. This core is removed, and all or part of it is taken as the sample. Simple versions of this type of sampler are inserted by hand and work well with fine-textured soils without too much moisture. Under such conditions it can be used to sample 1 meter or more in depth. Under favorable conditions this is the quickest and easiest type of sampler to use. This sampler is not productive when used with very dry, sandy soils, however. A

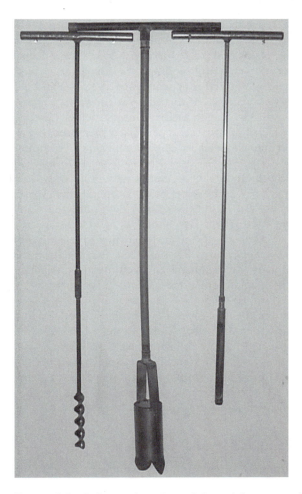

FIGURE 3.6 Soil samplers. From left to right—auger, bucket auger, and core-type samplers.

variation on this sampler is a special type in which the soil core is contained in a sleeve or liner that fits inside the sampler and can be removed with the sample inside. This type of sampler is particularly good for samples in which contamination or loss of a constituent is of concern. More sophisticated core samplers remove cores under even more controlled conditions (e.g., some will even remove a continuous core over the depth of drilling). Such cores can be important in reducing contamination and maintaining the physical and chemical integrity of the sample.

All three types of samplers can be power-driven, which allows deeper sampling. Powered samplers are frequently mounted on vehicles of various types (e.g., on a pickup truck or a special four-wheel-drive vehicle) and can sample several meters deep. (An example of this setup is shown in Chapter 5.) For very deep sampling a drill similar to or the same as the ones used to drill wells may be needed. Deciding on the type of sampler needed and obtaining the use of the proper sampler is essential in the presampling stage.

Often knowing the depth at which a sample is obtained will be important. For this situation a core sampler is best. Although this type of sampler will compress the sample it is still possible to determine the depth from which the sample came. With auger-type samplers soil from the sides of the hole will invariably be scraped off and included in the sample. For this reason these types of samplers are not preferred for situations in which the depth of the sample is important or in which a clear distinction between contaminated and uncontaminated sample is difficult to determine.

Samplers should be chosen for the type of soil or other material that is to be sampled and should be easy to use. Many samplers are easy to understand and use, while others are not, and sampling personnel must be taught how to use them (e.g., well-drilling rigs will require experienced drill rig crews to operate them). Studies have shown no difference, however, in analytical results that can be traced back to the type of sampler used (common soil samplers are shown in Figure 3.6) [6].

3.7.2. Powder and Granular Material Samplers

There is another type of sampler designed for dry powders and granular material. This instrument looks like a soil core sampler except that it has a solid point on the end and three compartments along the shaft. There is a sleeve on the outside so that it can be rotated to cover the compartments. The sleeve is rotated to cover the compartments and the sampler is inserted into the material to be sampled. Once inserted to the desired depth the sleeve is rotated so that the powder or granular material can flow into the compartments. The sleeve is once again rotated to cover the compartments and the sampler is withdrawn. This type of sampler is most often used to obtain samples in grain bins.

Using this type of sampler three samples are obtained at the same time, and usually several additional places in a grain truck or bin are sampled. Only dry powder and granular material can be sampled effectively using this sampler. Granules larger than a typical corn seed or moist, fine, wet, and very coarse materials are not effectively sampled. It could be used to sample coarse, dry sands under the proper conditions, however. A typical use of this type of sampler would be to obtain corn seed samples to check for foreign seeds or other contamination and moisture levels.

3.7.3. Water Sampling

Several types of water sampling are common. Surface waters and wells are sampled by simply filling a sampling bottle or with water pumped out of a well. Sampling bottles can simply be dipped into surface waters to obtain a sample. For pumped water it is common to clean the outlet, sometimes even to try to sterilize it, and allow some water to run for a period of time before taking the sample. Obtaining water from various depths is usually done with a sampler that is similar to the powder sampler described above. It will have a valve that works such that when it is at the required depth the valve is opened and a sample taken. The valve is then closed and the sample retrieved.

3.7.4. Sediment Samplers

The bottom of bodies of water is saturated with water. Taking a sample of this material requires a sampler that can retrieve both the solid and the water. The sampler bottom is open when it enters the sediment and is then closed to retrieve the sample. An even more complicated and specialized sampler is required when the sample pressure must be maintained [10].

3.7.5. Air Sampling

Air can be sampled using an evacuated sampling container. When the container is opened the surrounding air moves into the container. Other methods include pumping air into a container by various means. Pumping air always carries the risk of the pump contaminating the air sample. On the other hand, negative pressure can be used to move air into a container without contamination. Figure 3.7 shows a number of sample containers, including an air sample container, that can be evacuated.

3.8. OTHER SOIL SAMPLING CONSIDERATIONS

At this point the number of samples and the number of grids from which they should be obtained is not known. Only after the transect sampling has been done and the analytical results reviewed can this decision be made. There are some things that can be decided upon at this point, however.

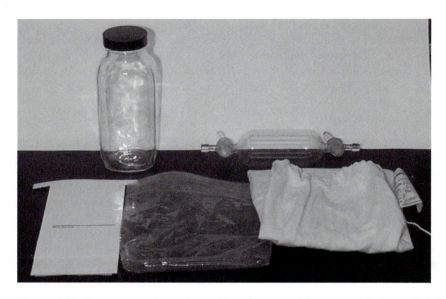

FIGURE 3.7 Sample containers. Front (from left to right)—paper, plastic, and cloth bags. Back (from left to right)—glass bottle and glass gas sample bottle.

When initially surveying the area and obtaining information (i.e., history) about a field, its pedology* and previous use, crop production, manufacturing, storage, and so forth should be gathered. The accuracy and precision of sampling is best when both of these factors are considered together in developing a sampling plan rather than when one is used alone.

The number of samples or cores taken, the size of the sampling pattern, and the depth of sampling are all important in obtaining accurate results. Three to six mixed cores give four to five times better analytical results than analyzing the cores individually. Smaller sampling grids give significantly better agreement between samples than do larger grids. This of course has practical limitations, particularly if the contaminated area or fields to be sampled are large. Also, if the problem is not agreement between samples but finding threshold values, this issue may be moot. Also, when sampling with depth recognizing the importance of horizonation is critical.

---

*Pedology is the scientific study of soils, including horizon formation, texture, and structure soil types.

3.9. SAFETY

At this time information is collected that is based on the initial survey and history and needed for safety in the field during sampling. Safety procedures, safety training, emergency numbers, and development of a chemical hygiene plan all need to be completed. Sources of safety clothing and equipment need to be located and appropriate orders placed for needed equipment. Arrangements also need to be made for the safe storage and transport of samples and equipment.

Sampling is not a casual activity. This is true even if there is no expectation of encountering dangerous chemicals. Long pants, long-sleeved shirts (with stomach covered), socks, boots, and hats are always required. Apparel covering less of the body is not acceptable during sampling. The two most important pieces of safety equipment for sampling are safety glasses or goggles and gas mask. Sufficient numbers of safety goggles must be available at all times. All persons present during sampling must wear safety goggles whether or not they are actually doing any sampling. Gas masks should be kept nearby at all times in case they are needed. If significant contamination is present, other appropriate apparel (e.g., coveralls, gloves, boots) also needs to be available.

If the contamination is dangerous, toxic, or corrosive, additional safety precautions, such as fully protective clothing (including coveralls, boots, full face shields, respirators, and gloves that are not only chemical-but also abrasion-resistant) must be used. Detailed arrangements need to be made for decontamination of workers and others visiting the field. Arrangements can thus be made at the field office for changing out of coveralls, boots, or boot covers when leaving. The field office should have a wash area for hands and face, including an eye wash station and an emergency shower in which workers massively covered with contamination can be washed. This is essential because such contamination must not be taken off the field and certainly not into a hospital!

All workers need to be constantly reminded to wash when leaving a contaminated field. This means both hands and face. (This assumes that these are the only two body surfaces exposed during sampling.) This also means that washing is required even if protective face shields and gloves have been worn, because hands pick up food, and contaminants on the hands can enter the body via the mouth [11].

3.10. SAMPLE CONTAINERS

Arrangements for obtaining a stock of the proper sample containers need to be made at this juncture. Sample containers can be constructed from glass, cloth, paper, and plastic in many different forms. Just as the sampler composition needs to be considered in sampling, so does the sample container. Sample containers must be compatible with the sample and not decompose, add to, or subtract from the sample components of interest. Many companies sell specially cleaned and treated sample containers. In some cases these are vials, which may come with a bar code containing a lot number and serial number already on them. Usually these are glass bottles or vials with screw tops, including a liner. The liner can be a simple rubber material or it may be a rubber backed by Teflon.

Bottles and vials with lined crimp-on caps are also common. Crimped tops require a crimping tool and a separate top and liner. A sample is placed in the container. A liner is added and then crimped in place. Such tops are not easily removed, so tampering is not possible and accidental loosening and spillage is prevented. Special crimped top removal tools are also readily available.

Both bottles and vials may have caps, either crimped or screw-on, with a hole in the center so that samples can be taken directly from the bottle using a syringe. Direct removal of a sample from the bottle without opening it prevents unintentional contamination. When applicable, these can be timesaving, as the bottles can be purchased to fit an automatic sampler.

Glass bottles have several drawbacks. First, they can be broken (although most small vials are quite robust). Second, if clear bottles are used sunlight can affect the sample contents, sometimes even changing the concentration of the component of interest. If samples are always protected from direct sunlight this is not a problem. If protection cannot be provided, however, brown bottles are preferred.

Other sample containers are made of cloth, paper, and plastic. All three of these can be quite porous. Cloth is the most porous of the containers and does not make good storage containers unless drying of the sample is particularly advantageous. Paper bags with or without plastic or plastic-coated liners are commonly used for soil samples. These bags are quite handy for many types of samples. Depending on their construction, drying of the sample may or may not take place.

Of the three common sample container materials, plastic is the least porous to water, but it can be very porous to organic vapors. In addition, plastics can react to or even be dissolved by organic compounds. Specialty plastic containers are available. The specifications of these containers must be investigated before their use, however. As with glass, clear plastics will allow light to interact with the sample if precautions are not taken to store them is a dark place. Most plastics can be punctured easily, and so must be handled carefully. Figure 3.7 shows some common types of sample containers. A summation of sample containers and their characteristics is given in Table 3.3.

TABLE 3.3 Characteristics of Sample Containers

Material	Characteristics	Good points	Deficiencies
Cloth	Strong, unaffected by water	Easy to handle and mark, strong	Contents dry, soil can sift out of bag
Paper	Inexpensive, adversely affected by water, may react with contaminants	Easy to handle, and write on	Soil will dry even in closed bag; no sifting; writing can smear
Plastic bags	Inexpensive, hard to write on, marking rubs off	Easy to handle, moisture retained	Porous to very porous for volatile organics and gases, subject to puncture, may react with contaminant
Plastic-lined cloth and paper	Similar to plastic except stronger and less subject to puncture	See plastic above	See plastic above
Special plastic	Specially treated plastic bags with Mylar or aluminum lining	Easy to handle	Not impervious to all organics (check manufacturer's specifications)
Rigid plastic	Shatter-proof; make sure container is chemically clean[a]	Easy to handle, nonwetting	Reactive with sample components; contamination of sample
Glass	Unreactive	Easily cleaned; can be dried in oven	Subject to breakage

[a]Plastics are often manufactured using oils to release them from such things as molds. Note that in some cases a certain type of container may be specified for a specific type of contaminant; otherwise, a container compatible with the contaminants expected in the sample is used.

In many cases the analytical laboratory chosen to do the analytical work will have specific bottles and other containers, along with the closures that it specifies for the type of sample being obtained and the analysis required. The laboratory will supply the bottles, along with directions for sample addition, amount, sealing, and returning to the laboratory.

3.11. TRANSPORTATION AND STORAGE

Arrangements for the transportation and storage of samples also needs to be decided on during this phase of the work. In many situations the analytical laboratory will have procedures that it requires for sample transportation and storage. These must be followed if the analytical results are to be meaningful. Note that different laboratories may have different procedures. The analytical results are produced assuming the procedures for storage and transport have been rigorously followed, however. Not following these procedures can adversely affect or even invalidate the analytical results altogether.

Once in their container, in the field the samples should be placed in a box for transport to the field office. A paper or wooden box is sufficient, although a polystyrene cooler may also be used effectively and is probably preferable, as it is both sturdy and insulated. The samples can be packed with paper, packing peanuts, or bubble wrap, or if cooling is required ice can be used as a packing material for protection, insulation, and temperature control. When full, the box is transported to the field office for storage. A separate storage room—or, preferably, building—is needed for the samples. This will prevent contamination of the work space with material from the samples and prevent contamination of the samples from the work space. At this point the samples are ready to be either analyzed at the field office (sample characteristics such as pH are analyzed in the field as soon after sampling as possible) or shipped to the analytical laboratory.

Many restrictions apply to shipping dangerous or toxic compounds. All applicable safety precautions, signage, and other shipping restrictions must be followed when shipping samples to an analytical laboratory.

3.12. CHAIN OF CUSTODY

From obtaining a sample through the final analytical procedure applied to it, a chain of custody must be kept. This is a trail or list of where the sample has been, who has handled it and how, the length of time it was stored, and its transportation route. This information must be meticulously recorded and maintained because it provides proof of the authenticity of the sample.

It also helps in preventing contamination of the sample, or if analysis indicates that something untoward has happened between sampling and analysis the chain of custody, can shed light on where and when this might have happened. Chain of custody forms may be available from the commercial laboratory that is to analyze the samples. More information about the chain of custody will be given in Chapters 8 and 10.

3.13. ANALYTICAL LABORATORY AND PROCEDURES

Samples will be analyzed at two locations for one or many components. Analysis that must be carried out immediately will be performed in the field or at the field office laboratory discussed above. Samples needing more detailed or sophisticated analysis will be sent to a commercial laboratory. At the commercial laboratory samples will be treated appropriately for the analysis of the component of interest.

An analytical laboratory needs to be decided upon and at the same time the types of components the laboratory will be asked to analyze for need to be decided. The commercial laboratory will have a project manager assigned to the project. This person will help in making decisions about various aspects of sampling, sample handling, shipping, and the analytical methods needed to obtain the desired information. Often the analytical laboratory will have special kits with appropriate sample containers and shipping boxes, which may vary, depending on the type of analysis to be done. They may also specify the conditions such as temperature needed during storage and transport. These details are further discussed in Chapters 8 and 10.

The question then becomes what components in the sample are important. The first most important division is between organic and inorganic. Samples to be analyzed for inorganic components are obtained and handled differently from samples intended for the analysis of organic components. As noted above the metal composition of the sampler will be important in obtaining samples to be analyzed for inorganic components, particularly metals. Samples to be analyzed for organic components must be handled in such a way as to minimize the loss of organic compounds.

If the laboratory does not have sample shipping kits, discuss with them how the samples should be handled. Along with this goes deciding on a numbering system for samples and setting up the chain of custody, which must be decided upon and well described and established. All of this is also related back to the project notebook and the question of the use of bar codes.

3.14. STATISTICS

All sampling and analysis will have variation, which is normal and expected. If variation does not occur it is reasonable to suspect that the data are in error. Because of this variation it will be essential to analyze the analytical results using statistics. These tools will be used to be determine if samples taken at different places and times have different amounts of a particular component or not. In many cases it will be important to know if the results are within acceptable limits or not. Statistical analysis will also be important to determine if an analytical result is an outlier or in error. Statistical analysis of analytical results thus will be essential.

What statistical tools will be needed? The first calculation to be made is the mean and standard deviation of the samples. Subsequently a t-test will be needed to determine if two samples are different or the same or if the sample is within acceptable limits. This may be between two samples from the same or similar places in the field or samples obtained at different times. In all probability some more advanced geostatistical tools will be needed to graphically represent the level of components of interest. These representations will be used to understand what is happening in the field. They will also be needed to explain the sampling needs to others, including the public.

Because there are many statistical tools and packages, some evaluation of the various packages may be needed. For the statistical tools and programs to be most effective personnel must know how to use them before sampling begins. Often the more powerful programs will require more learning before they can be used effectively, and thus it may be more useful to use a statistical program that the operator knows how to use rather than to try to learn a new program. Another thing to remember is that the tools chosen must be compatible with the computer system to be used.

In all cases it is essential to have a statistician check the statistical analysis for errors and to make sure the correct tools have been used.

3.15. OTHER TOOLS TO CONSIDER

The use to be made of other sampling tools needs to be decided upon. How much use will be made of GPS? What type of receiver will be needed? Is GPR called for? How much or how will remote sensing and GIS be used? If these tools are to be used, what computer software will be needed and what computer system will be used to collect and store the data?

A word of caution would be that some of these applications use large amounts of computer storage space. When deciding on a computer, both RAM and hard disk storage should be maximized, thus RAM of 512 MB and an 80-GB hard drive are the minimum recommended. Removable storage for presentations and backup must also be provided. Either a CD or DVD recorder will be best suited to this task. Because CDs are inexpensive (less than 6¢ apiece), CD-R (not CD-RW) CDs are recommended. These are record-only, so they make a permanent record of the work done and presentations made.

3.15.1. GPS

The use of GPS for locating sampling sites in the field is highly recommended, and in many situations will be essential. Sample sites can readily be located even if other markers are lost or destroyed. A GPS receiver can be set up to locate preselected sample sites in a field, and will not only tell the person sampling when he or she is at the correct location but how to get from that site to the next. In this way relatively unskilled personnel can be used in sampling. The movement of water and other components of interest can be followed by noting their change in position. The rate of movement can also be obtained from the data, either directly, allowing the receiving instrument to do the calculation, or indirectly, by noting the change of distance and the time. More discussion about this use will be given in Chapter 11.

Decisions as to the type of receiving units needed are made at this point. Such questions as the sophistication needed and whether a handheld or vehicle-mounted unit are to be used must be answered. The equipment, including any software needed, should be bought, and operators or persons doing the sampling should be trained in their use [12].

Knowing the position of an airplane or boat far from landing or shore, or a sampling site in fields that are hundreds or thousands of hectares in size to a precision of $\pm$ 15 meters is sufficient. When sampling a field of 1 ha, however, knowing a sampling site to $\pm$ 15 meters is not good enough. There are two ways to improve the accuracy. The first is to use position averaging. A receiver capable of position averaging takes many readings while stationary and averages them. In this way the position to $\pm$ 3 meters can be obtained. This may be sufficient for many sampling situations. Two other ways to obtain more precise positions are to use either a wide area augmentation system (WAAS) or a differential GPS (DGPS)-capable receiver.

The WAAS methodology uses ground stations to correct signals received from satellites, and a corrected signal is then broadcast by satellites. When a receiver capable of accessing the WAAS system uses this information the variability in the accuracy of a site position is less than $\pm$ 3 meters. The Federal Aviation Administration (FAA) and the Department of Transportation (DOT) are developing this system, which is available without charge to any GPS unit capable of receiving the signal and making use of it. The WAAS is limited to North America and to areas that do not have obstructions that interfere with the signal.

The second way is to use DGPS. It is similar to WAAS in that it uses corrected signals. A receiver capable of DGPS is necessary. The use of DPGS can increase the accuracy to ± 2 cm. Obtaining DGPS accuracy using civilian GPS receivers however, requires that a DGPS site be within range. The DGPS site is an accurately known location that broadcasts position correction information to GPS units. This information can be used in real time or after the fact to find the exact location. Real time would be preferable when the exact place in the field needs to be sampled several times. Postprocessing, finding the location after taking the sample, and returning to the field office, can also be done and is less expensive. If taking a sample from close to a previously sampled site is sufficient, postsampling DGPS is sufficient. With this capability sampling sites can be determined with exceptional accuracy.

Global positioning systems and equipment are constantly being upgraded and improved, thus when deciding on a GPS receiving unit to use, all the latest advancements in this technology should be investigated, along with best estimates of where the technology will be in the foreseeable future.

3.15.2. Ground-Penetrating Radar

Ground-penetrating radar is called for when the field history indicates that something might be buried in the field or when the field has been traversed by armed forces. The field history is particularly important in determining if something dangerous, such as explosives, might be buried in the area. Ground-penetrating radar is also indicated if there is a likelihood that there are buried tanks, cables, pipelines, or drainage tiles. A hole 1 meter in diameter and $1^1/_2$ meter deep was found in a field in which a drainage tile had broken and soil had been washed in. (Such a situation is called a blowout, even though soil has been eroded into the tile.) This type of field condition can be hard to see until one is nearly in it. It is thus extremely dangerous for personnel and equipment. Ground-penetrating radar can show where such blowouts are developing. It can also tell where drainage tiles are so that they can be avoided during sampling.

Ground-penetrating radar could also be used if the area is suspected of being an abandoned village or city [13]. In this case it would help determine areas under which tunnels and caves are located, and thus needs special attention when sampling. Tunnels and caves can collapse during sampling and trap and even kill personnel, and they thus represent a hazard to sampling personnel and equipment.

3.15.3. Remote Sensing

Remote sensing in the sense of pictures taken from airplanes or satellites can be valuable in locating underground features. These are often visible as differences in either the type or the growth of vegetation. Such pictures can help to determine the sampling pattern that will be the most useful in a particular field. It can also be useful in showing the differences in the field before and after the sampling and remediation activities [14].

There is a great wealth of remote sensing information available from the U.S. government. Much of this comes from NASA's Earth Observing System (EOS). There is also an Earth Resources Observation Systems (EROS) data center, which specializes in land processes. Other areas of the environment are covered by other organizations. All these organizations and their data are brought together in the Distributed Active Archive Center (DAAC). Much of this information is free or available at a nominal cost [14–17].

3.15.4. GIS

Will a GIS system be used and how will it be used? It is a powerful system that can be very helpful in relating data from different sources. Water movement through soil combined with topography, soil type, and contaminant concentration can be combined on one map of the field. This combined picture produces an informative map of the area [18]. The most common computer software used for GIS is called ArcView. There is other software that will do the same things, and all should be investigated before deciding on a system to use. Note that even if you do not intend to start your own GIS you may need some or all of the software to view and work with GIS produced by local, state, and national governments.

3.16. MODELING

Data collected during sampling, or obtained by analysis of samples or from other sources will be useful in building a model of the field and how it will react in the future. We can think of this as a model of the field before sampling is started. It can also be a dynamic model that will allow calculation of the expected future characteristics of the field. These models can be very effective in explaining what is happening, in deciding where to sample, what to analyze a sample for, and what type of follow-up sampling will be needed.

Modeling requires a model. This can be a simple physical model or a complex computer simulation. A physical model can provide information about many aspects of the field but will have limited predictive capability. A complex computer simulation can be used to predict the course of the work and will also allow changes in the inputs to explore various situations that may occur in the future. It will also allow changes in the model as more data become available.

Modeling is another component of the sampling plan that needs to be explored at this time.

3.17. PERMITTING

In many instances it will be necessary to have permits for a sampling and remediation plan. These will generally not be needed when sampling farm fields for such things as plant nutrients. Sampling for toxic or dangerous materials or other situations may require permits, however. Permits can be of many different kinds; earth disturbance, erosion and sedimentation control, water obstruction and encroachment, and various permits associated with landfills are a few examples of activities that may require a permit. Permitting is a state activity, thus different states require different permits. One state may require a permit for an activity while another does not. It is best to go to the state or local Environmental Protection Agency office and check to see what may be needed in this regard.

3.18. RESOURCES

Commercial and governmental sources of the equipment, supplies, computer software, and so on, mentioned in this chapter are given in Appendix B. Many times the place to start the search for companies is at their Web site.

3.19. CONCLUSIONS

Once all the above information is obtained and all the decisions are made it is still not time to start sampling. The needed equipment must be purchased and brought to the field office. A laboratory and storage area must be prepared. Sample handling and transportation procedures need to be written down, and a chain of custody protocol developed and appropriate procedures for its implementation decided upon. With the above information a detailed safety plan for sampling the field can be prepared. No sampling can begin until all the safety precautions and equipment have been obtained and everyone knows how to use them.

At this point a preliminary sampling plan can be put together. It should include plans for transact sampling and detailed sampling. (See Chapter 5.) It is good to keep in mind that during most transect sampling too few samples are taken. For this reason one should err on the side of too many samples during transect or preliminary sampling.

QUESTIONS

1. Describe the characteristics of a project notebook.
2. What types of information can be obtained by walking around the outside of a field in relationship to the sampling to be done?
3. Make a list of the types of information that one should obtain about a field when preparing a summary of its history.
4. What two soil characteristics may have a pronounced effect on the type of sampler used in sampling a field?
5. What three basic types of samplers are there? Explain where each might be used most effectively.
6. What two pieces of safety equipment are most important to have in the field and at the sampling site?
7. What kinds of statistical tests of data will be needed in analyzing sampling and analytical results?
8. Explain why GPS is essential in obtaining accurate and repeatable sampling results.
9. Explain the following abbreviations. (Do not hesitate to use other chapters in your explanation.)
   (a) GPS
   (b) GPR
   (c) GIS
   (d) EROS
   (e) Remote sensing
10. Describe the characteristics of a field office, laboratory, and storage area. Why is it a good idea to have the sample storage area separate from other rooms?
11. What advantages do DOQQs have over other types of photographs or maps?

REFERENCES

1. Kanare HM. Writing the Laboratory Notebook. Washington, DC: American Chemical Society, 1985.
2. Palmer RC. The Bar Code Book: Reading, Printing, and Specification of Bar Code Symbols. 2nd ed. Peterborough, NH: Helmers, 1991.
3. National Mapping Information—Digital Orthophoto Program. United States Geological Survey. http://www.usgs.gov/. Look for DOQQ.
4. Ramos B, Miller S, Korfmacher K. Implementation of a geographic information system in the chemistry laboratory: An exercise in integrating environmental analysis and assessment. J Chem Ed 2003; 80(1):50–53.
5. MrSID GeoViewer. Lizardtech software. http://www.lizardtech.com/download/.
6. Wagner G, Desaules A, Huntau H, Theocharopoulos, S, Quevauviller P. Harmonisation and quality assurance in pre-analytical steps of soil contamination studies—Conclusions and recommendations of the CEEM soil project. Sci Total Environ 2001; 264:103–117.
7. The Use of Historical Data in Natural Hazard Assessments. Glade T, Albini P, Francés F, eds. Boston: Kluwer Academic, 2001.
8. Huang CC, O'Connell M. Recent land-use and soil erosion history within a small catchment in Connemara, Western Ireland: Evidence for lake sediments and documentary sources. CANTENA 2000; 41:293–335.
9. Sastre J, Vidal M, Rauret G, Sauras T. A soil sampling strategy for mapping trace element concentrations in a test area. Sci Total Environ 2001; 264:141–152.
10. Fisher MM, Brenner M, Reddy KR. A simple, inexpensive piston corer for collecting undisturbed sediment/water interface profiles. J Paleolim 1992; 7:157–161.
11. Reynolds EM, Randle Q. Pocket Guide to Safety Essentials. Washington, DC: National Safety Council, 2002.
12. Letham L. GPS Made Easy. 12th ed. Seattle: Mountaineers, 2001.
13. Conyers LB, Goodman D. Ground–Penetrating Radar: An Introduction for Archaeologists. New York: Rowman & Littlefield, 1997.
14. Lunetta RS, Elvidge CD. Remote Sensing Change Detection: Environmental Monitoring Methods and Applications. Chelsea: Ann Arbor Press, 1998.
15. Earth Observing System. NASA—Goddard Space Flight Center. http://eos.gsfc.nasa.gov/.
16. Earth Resource Observation Systems (EROS) Data Center. USGS. http://edcwww.cr.usgs.gov/.
17. GSFC Earth Sciences (GES) Distributed Active Archive Center. NASA. http://xtreme.gsfc.nasa.gov/.
18. Korte GB. The GIS Book. 5th ed. Albany, NY: OnWord Press, 2001.

4

Safety

Field sampling is not a leisure time activity.
Safety must be practiced at all times.
Safety equipment must be used.
Clothing must be appropriate to the hazard.

A safe working environment is not an accident; it requires thought and work. Safety is good from human, business, and economic perspectives. To realize these benefits a number of things need to be done. First, safety information and physical resources are identified or found. The types of fields to be sampled must be identified, along with the type and concentration of the contaminants they contain. Once this is known, personnel protection can be planned. Additionally, off-field and environmental protection must be considered. When this is all brought together, a safety plan including the use of a chemical hygiene plan (CHP), material safety data sheets (MSDS), and other resources such as the *Merck Index* can be incorporated.

Safety can be, and often is, viewed as an unwanted expense. In reality it always results in enormous savings. One key worker not doing his or her job for half a day can cause serious delays in completing a sampling exercise. Often an accident happening to only one worker will involve other workers as they help care for or evacuate the injured person. In this case the work output of several persons is adversely affected. Add to this the cost of transportation and medical treatment, and the expense can be quite a burden.

Even if everything is covered by insurance, repeated use of the coverage will result in the increasing cost of insurance. From all ways of assessing accidents they are always costly. This is a cost that need not be borne if personnel observe proper safety procedures and precautions while on the job.

With all this said, "Who actually suffers when an accident occurs? The answer is always the injured worker. The company suffers because of the lost work time, the need to replace the worker, the increase in insurance costs. But, truly, in the end the only one who really suffers is the injured worker. They suffer due to the injury itself, the recovery time, the possible loss of a limb, the loss of wages, the suffering the family has to endure, and the possible loss of the job. For this reason workers need to always use the proper personal protective equipment (PPE) because they are the most important reason to protect themselves from possible injury" [S. Ullom, Ullon Safety Resources, Inc., personal communication. (See Appendix B.)]

4.1. AREAS OF SAFETY CONCERN

There are five areas of safety to consider in sampling a contaminated field. The first consideration is the safety of the personnel doing the sampling. What kinds of hazards will personnel likely be exposed to? Will the fields contain fumes from gases, liquids, or solids? Will they encounter flammable, toxic, or corrosive liquids? Will dangerous solids be present? Answering these questions will allow for the development of good safety measures. The second part of this is to ask, "Do the personnel have the training and expertise to be in the field and work with the components there, including sampling equipment, without contaminating themselves and others?"

A second essential component in answering the above questions is to consider the level of contamination likely to be encountered. Areas containing high levels of contamination are more dangerous than less contaminated areas. These areas require more exacting safety precautions, and equipment and safety must be more carefully thought out. In no case, however, no matter how safe or how uncontaminated a field seems, should safety be taken for granted.

The United States Occupational Safety and Health Administration (OSHA) standard 29 CRF 1910.132 contains a requirement for a prework personal protection equipment (PPE) assessment. This consists of filling out a personal protective hazard assessment worksheet, such as that shown in Table 4.1. This sheet needs to be filled out and signed by the person doing the sampling to make sure that he or she fully understands the safety requirements. Note that not all hazards and not all PPE will be needed with

TABLE 4.1 An Example of a Hazard and Personal Protection Worksheet

Hazard and PPE* Assessment Work Sheet				
Field Location :				
Date, Name and Signature of Individual Completing Assessment				
Anticipated Hazards Associated with Sampling. Check all that apply.				
[] Chemical	[] Compression	[] Respiration	[] Impact	[] Atmospheric
[] Splash	[] Penetration	[] Noise	[] Light(optical)	[] Electrical
[] Engulfment	[] Laceration	[] Heat/Cold	[] Radiation	[] Bloodborne
All PPE* required for safe sampling				
Head				
Eye				
Ear				
Respiratory				
Hand				
Clothing				
Chemical Protective Clothing				
Foot				
Electrical				
Other				
Comments				
* Personal Protective Equipment				

all sampling environments. A thorough assessment of hazard and protection needs for each field must be done, however.

The third consideration is the field itself and the people, animals, and plants around it. Is there a possibility that sampling will cause any movement of contamination off the field? This might be as solids tracked out by wheeled vehicles or as dust brought into the air by various sampling activities.

The fourth is to make sure that contamination cannot move off the site during nonworking hours. The field must be protected from water and wind erosion. Also, are underground areas and water supplies safe from contamination? Will sampling activities allow contamination to move into uncontaminated areas? The fifth is the safety of the personnel transporting and analyzing the samples obtained. If samples are in appropriate containers contamination of people and their surroundings during transportation should not occur. Laboratory personnel must be apprised of the expected contaminants in the samples, however, so that appropriate safety precautions can be taken.

4.2. SAFETY RESOURCES

Information about safety in general and in sampling situations can be found in various locations. Often topics relating to safety in environmental work are referred to as environmental health and safety (either as EHS or EH&S) [1]. These are key words when investigating specific safety concerns. The National Safety Council Web site has many publications that can be helpful in any situation [2]. OSHA is another good source of safety information, along with information about chemical hygiene plans [3]. The U.S. Environmental Protection Agency (USEPA) also has information about safety, particularly in terms of environmental contamination [4]. Safety information in relationship to the dangers of a particular chemical can be found in material safety data sheets, better known as MSDS sheets [5]. Additional safety information, including use and toxicity information about chemicals, can be found in the *Merck Index* [6].

There are consultants and consulting organizations that can help with developing safe working conditions. In addition they can help in making sure that all government organization safety rules are followed. This can include ten 30-hr safety courses, which are also known as outreach training programs and the 40-hr (HAZWOPER) courses associated with safety and safe sampling issues. Examples of the types of topics covered in these courses are given in Table 4.2. In addition to these programs most states

TABLE 4.2 Examples of Topics Covered in the 10-, 30-, and 40-Hour Courses

10-hr course	30-hr course	40-hr course
Walking and working surfaces	Walking and working surfaces	Principles of safety
Means of egress	Means of egress	Planning and organization
Fire protection	Fire protection	Preparation for field work
Material handling	Hazardous materials	Chemical hazards toxicology
Electrical, lockout/tagout	Personal protective equipment	Monitoring instrumentation, confined space
Machine guarding	Confined spaces	Trenching and excavation
Confined space	Materials handling	Medical surveillance
Bloodborne pathogens	Electrical machine guarding lockout/tagout	Personal protective equipment, levels of protection
Hazardous materials	Welding, cutting, brazing	Fire protection
Personal protective equipment	Pathogens	Decontamination procedures
Record keeping	Hazard communication	Drums/containers sampling and packaging
Hazard communication	Record keeping Safety and health	Site emergencies

Note: The 40-hr course will only be found designated as Hazwoper. These are general topics covered and are not intended to represent any actual course content.
Source: Ref. 2.

have free safety consultation programs, a list of which can be found at www.OSHA.gov.

While these are excellent sources of information, nothing can replace careful thought and thorough preparation when working in hazardous situations. For example, spilling a quantity of liquefied gas, even nitrogen, can quickly fill a room or enclosed space with that gas and thus cause suffocation. If the gas is flammable an explosive situation can also quickly develop. For this reason knowing only the toxicity or the flammability of a gas is not enough. The question becomes what will happen if a spill occurs and what needs to be done to protect people if this happens.

Another example would be spilling liquids. A volatile liquid spill can cause the same hazards as described above; that is, the fumes may be flammable and can thus cause an explosion if a source of ignition is present. On the other hand, spilling an acid may cause damage to the surface on which it is spilled, produce no fumes, and cause no other harm. Some acids and bases, however, such as hydrochloric, nitric, and acetic acids and ammonium hydroxide can release fumes that can be overpowering. If acid is spilled on a reactive metal, hydrogen gas can be released and cause a dangerous or even an explosive situation to develop.

4.3. BASIC SAFETY

At the very start of a project all personnel need to be instructed in basic safety procedures and practices. This may mean a basic Red Cross first aid course or in-house instruction, including cardiopulmonary resuscitation (CPR) instruction. First aid supplies should provide gloves and CPR shields to protect first aid givers from bloodborne pathogens. Basically all personnel need to know first aid because it is not possible to know beforehand who will be the person needing help [7]. General safety rules for all situations are given in Table 4.3 [8].

Along with this table should go emergency phone numbers. These must include life squad, fire department, hospital, hazmat, sheriff, police, state EPA, and other appropriate entities. Communications must be set up to facilitate contacting the appropriate organization without delay. Every person involved in field sampling must have easy access to such communication.

4.4. SIGNAGE

Safety signs relating to safety concerns in and around the field and field office, laboratory, and storage areas must be installed. It is not sufficient to simply have signs placed in appropriate areas; workers also need to know

TABLE 4.3 General Safety Rules

Know basic first aid.
Do not sample alone.
Assess and select the proper PPE.
Know what types of materials are being sampled.
Have safety equipment, including fire extinguishers, and first aid nearby.
Know where help is and how to get you to it and it to you.
Store samples appropriately and away from food.

how to read the signs. In Figure 4.1 the sign illustrated at the top left is an example of a hazard warning label. This indicates the general type of hazard present.

The label on the top right is an example of a hazardous material identification system label (HMIS). This label not only provides the types of hazards involved but also the severity of the hazard. The numbers can range from 0 to 4, indicating an increasing level of hazard in each of the four small diamonds. The specific type of hazard is indicated by both the position and the color of the interior diamond: the upper left diamond is blue and is an indicator of the health hazard [3]; the upper right is red and is fire hazard [2]; the lower right is yellow and indicates the reactivity; and the lower white

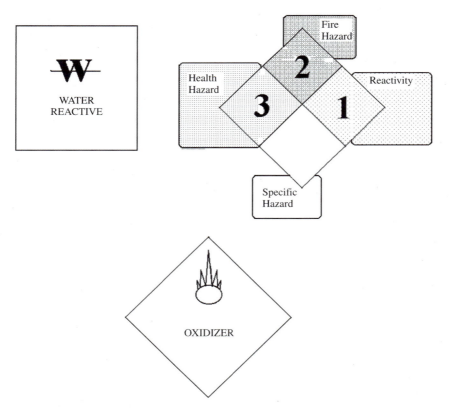

FIGURE 4.1 Examples of safety signs. Upper left: hazard warning label; upper right: hazardous material system label (see also Figure 4.2), and bottom: shipping label. Next page: explanation of label.

Explanation of Hazardous Material System Label (cont.)	
Diamond	Hazard danger designations
Health hazard—Blue (#3)	1—None
	2—Mild
	3—Extreme
	4—Deadly
Fire hazard—Red (#2)	4. Flash point below 23°C (73°F)
	3. Flash point below 38°C (100°F)
	2. Flash point below 93°C (200°F)
	1. Flash point above 93°C (200°F)
	0. Does not burn
Reactivity—Yellow (#1)	4. May detonate
	3. Shock and heat may detonate
	2. Violent chemical change
	1. Unstable when heated
	0. Stable
Specific hazard	Oxidizer—OXY
	Acid—ACID
	Alkali—ALK
	Corrosive—CRO
	Water reactive—W
	Radiation— ☢

FIGURE 4.1 Continued.

diamond contains special hazard information as needed. Also given is specific information contained in each diamond. Figure 4.2 shows another HMIS and its associated safety labeling.

The bottom sign on p. 97 is an example of a shipping label. These are typically seen on the sides of trucks and the outsides of shipping boxes. In this case the label tells the handler that the material in the container is an oxidizer. If field samples have some particular component that would be of concern if the package were dropped or if spilling the contents would cause a hazardous situation, there should be a label on the box alerting people handling the box.

Weather emergency plans and signs also need to be taken into consideration. Signs giving information about weather emergency signals, emergency exits, and routes to safe locations need to be in place. All workers must be aware of these signs and their meaning. The location of safe areas must be easy to determine and get to.

There are many other types of safety signs. Some of these are used to show the location of safety equipment, including safety showers, eyewash stations, safety equipment, and danger zones. It is advisable to make liberal use of safety signs [2].

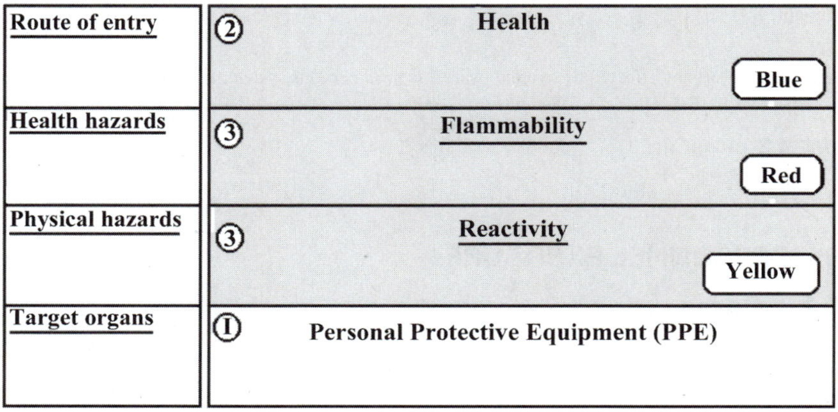

Route of entry	②	Health		
				Blue
Health hazards	③	Flammability		
				Red
Physical hazards	③	Reactivity		
				Yellow
Target organs	①	Personal Protective Equipment (PPE)		

FIGURE 4.2 Another hazardous material identification system. Shades of gray indicate background colors.

Key for Personal Protective Equipment	
A. Safety glasses	G. Safety glasses, gloves, vapor respirator
B. Safety glasses, gloves	H. Splash goggles, gloves, chemical apron, vapor respirator
C. Safety glasses, gloves, chemical apron	I. Safety glasses, gloves, dust and vapor respirator
D. Face shield, gloves, dust respirator	J. Splash goggles, gloves, chemical apron, dust and vapor respirator
E. Safety glasses, gloves, dust respirator	K. Air line, hood or mask, gloves, full chemical suit, boots
F. Safety glasses, gloves, chemical apron, dust respirator	X. Ask supervisor

Additional information

Route of entry—entry of chemical into body—ingestion, inhalation, absorption

Health hazard—effects of overexposure—acutely toxic, corrosive, or irritant

Physical hazard—characteristics which make chemical hazardous—explosive, oxidizer flammable

Target organs—organs affected by overexposure—central nervous system, liver, kidneys

FIGURE 4.2 Continued.

4.5. PERSONNEL EXPOSURE

4.5.1. Gases

Both compressed and liquefied gases are commonly found in the field. Compressed gases, such as nitrogen, oxygen, acetylene, propane, and helium, are common in a variety of agricultural and manufacturing facilities. If the top of one of these tanks is broken off, the tank turns into a missile capable of penetrating cement block walls. For this reason protective caps are placed on the tank during storage and transport. The tanks are secured to some substantial cart or building structure when in use. Some gases, such as acetylene, can ignite when escaping from a metal orifice and thus are of particular concern. Acetylene and propane escaping from a tank can produce a flammable and potentially explosive situation. Oxygen supports combustion, which means that both escaping oxygen or oxygen-rich environments will support oxidation and burning where otherwise it might not occur. Even inert gases such as nitrogen and helium can produce dangerous situations. When released in large volumes these gases can replace all the oxygen in an enclosed area, causing suffixation. Compressed gas tanks encountered during field sampling must be treated with extreme caution.

Some gases, including those discussed above, are commonly liquefied under pressure and are stored and transported in this form. Some are kept in insulated containers, while others are in simple pressure tanks. These liquefied gases pose some of the same risks as compressed gases in cylinders, and although they are not commonly under as much pressure as compressed gas, dangerous leaks can develop in the tanks and cause serious safety risks. A different hazard occurs when these liquids are spilled. One mole of liquid nitrogen, 14 g, will occupy a volume of 22.4 liters when it warms up and becomes a gas. The liquid can rapidly change to a gas, replacing oxygen and suffocating any people and animals present.

In some cases liquefied gases are a fire hazard, and some are toxic or otherwise hazardous. Flammable gases are colorless and must build up to a certain level before they can burn or more likely explode. Liquid propane will be converted to a gas when pressure is released and it will produce an explosive mixture with air. Unfortunately once these gases start burning, the fire is hard to put out. Liquid ammonia will become a cloud of ammonia gas, which is choking and toxic and will overcome a person trapped in the cloud. All visible gases—that is, gases that have color—are toxic; however, the reverse is not true. Many colorless gases can be toxic, however.

Fumes originating from a monitoring well can also be dangerous. They may be toxic or flammable, and so appropriate precautions must be taken at all times when opening a monitoring well.

4.5.2. Liquids

Liquids can either be pure organic liquids, mixtures of organic liquids, inorganic liquids, and solutions of organic and inorganic compounds in water. These liquids may be volatile or nonvolatile, and can release strong fumes and/or be caustic. Volatile liquids can fill an area with fumes that can burn. In this case the fire will be fed from the pool of liquid, and so covering the pool with an appropriate flame extinguisher can stop the fire.

Fumes originating from a volatile liquid are heaver than air and can fill and travel through low areas. If they contact a source of ignition, the flame can travel back to the source of the fumes and cause a large fire or explosion. When confined in a small space they can cause the suffocation of persons entering these spaces.

Many liquids present hazards of other types. Acids and bases frequently come as solutions of gases in water. Most common are the mineral acids—hydrochloric, nitric, and sulfuric—and the base ammonium hydroxide. When concentrated solutions of these are opened hydrogen chloride, nitrogen oxides, sulfur oxides, and ammonia gas are released. Because of this spills of these concentrated acids and bases must be treated with extreme caution. In addition to the fumes they contain, these acids and bases can react to release both toxic and flammable fumes. Diluted strong acids and bases often do not have a strong smell but must still be treated with caution, although they are not as dangerous.

There are many other caustic and/or volatile and flammable liquids and solutions that can be dangerous and thus must be handled with caution. Organic amines are bases and have a strong smell of fish. All toxic organic amines must be treated as if they can be absorbed through the skin, because many of them can. It cannot be guaranteed that amines will not penetrate rubber or synthetic gloves. (More information about gloves is given below.) For this and general safety precaution reasons amines should always be treated with the utmost caution.

Some inorganic and organic liquids and solutions are unusually dangerous. Some, such as solutions containing hydrides, will spontaneously ignite on exposure to air. Some organic solvents, such as diethyl ether (often called ether), will ignite if it is dropped on a hot hotplate. (No spark is necessary.) Under anaerobic conditions methyl mercury may be produced, which can pass through protective gloves and cause death. There are many other examples that people doing field sampling need to be aware of. This is why a history of the field is so important.

4.5.3. Solids

Often solids are easier to deal with because they can simply be picked up, put in appropriate containers, and then safely handled. There are numerous solid chemicals, however, that produce dangerous situations. Phosphorus spontaneously ignites and burns when exposed to air, and the fire is very hard to extinguish. Alkali metals such as sodium react with water, releasing hydrogen. The reaction is so vigorous that sparks are produced, setting the hydrogen on fire. Metal hydrides, which can come as solids as well as the solutions mentioned above, release hydrogen when they become wet and can produce explosions.

4.5.4. Unknown Materials

Extreme caution in handling any kind of unknown material is always prudent. When sampling a field it is also always essential to follow all safety precautions because it is not possible to tell when a sampler, handheld or machine-driven, might encounter and break open a container of hazardous chemicals. This can happen in an area that apparently has never had any waste dumped on it and is far from any population center. It can also happen so quickly that fumes can overcome persons in the area.

Dangers associated with all of the above potential contaminants are compiled in Table 4.4.

TABLE 4.4   Sources of Personnel Contamination

Gases	Permanent gases—may be toxic, flammable, explosive, or all three.
Compressed gas	Gas can escape, sometimes with autoignition, slowly or explosively from compressed gas containers. Gases may be toxic, explosive, flammable, or all three.
Organic liquids	May be toxic or produce flammable gas or explosive situations. All three can be produced at the same time, especially when the liquid is heated. Hot metal can cause ignition. Some organic compounds can be absorbed through the skin.
Inorganic liquids	The most common inorganic liquid is mercury, which is toxic. It and its compounds may be absorbed through the skin. Mercury fumes are also toxic.
Inorganic liquids and solutions	Containers of concentrated solutions may release gas when opened. Contact of solutions with other chemicals may release toxic, explosive, or flammable gases or all three. Contact with skin can cause burns.
Organic and inorganic solids	Can sublimate, releasing flammable, toxic, explosive gases. May be absorbed through the skin.
Unknown gases, liquids, and solids	Unless otherwise known, all unknown materials encountered are considered to fall into one of the above categories and may be a mixture of two or more of them.

4.6. PERSONNEL PROTECTION

Sampling a field involves entering it with personnel and equipment, and both need to be protected from contamination while the sampling is being done. Once sampling is completed personnel and equipment exiting the site need to be decontaminated [9]. This means washing, and thus an appropriate area needs to be constructed for cleaning both personnel and contaminated clothing and safety equipment. Washwater and other contaminated materials need to be cleaned or disposed of in an appropriate manner to prevent the spread of contamination outside the field. This is especially true for shoes or shoe coverings.

Appropriate safety equipment must be available at both the field office and in the field. Table 4.5 provides a list of common safety equipment that should be available. In addition to these items there may be specialized safety equipment needed for unique sampling situations. Careful thought about the sampling situation and the types of materials to be handled will be needed to provide a safe working environment and to have all potentially needed safety materials readily at hand.

TABLE 4.5 Standard Safety Equipment

Item	Description
First aid kit	Bandages (including liquid bandages), gauze pads, burn ointment, rubber gloves, antiseptic, tape, and items to prevent exposure to bloodborne pathogens
Eyewash	An eyewash station to provide 15-min flushing and portable eyewash bottles
Eye protection	Safety glasses and goggles
Ear protection	Noise-suppressing ear plugs—a minimum noise reduction provided by the ear protector must be able to reduce noise to the ear to 85 decibels
Breathing	Gas mask with assorted cartridges
Head protection	Hard hat
Gloves	Both work and rubber or chemical-resistant gloves
Water	Enough water to allow for emergence washing of exposed body areas
Personal monitors	Air monitors for exposure to fumes
Oxygen	Compressed tank of oxygen

4.6.1. Do Not Sample Alone

Do not sample alone. No matter how safe a field looks, if you are not familiar with it it is unsafe to sample alone. In addition to soft areas and unseen holes in the field there is the danger of stumbling onto an area containing noxious fumes. Sampling or monitoring wells can also often contain unpleasant surprises when uncovered. There is also the possibility of encountering animals, such as poisonous snakes, which can be quite dangerous. Sampling in pairs can be faster than sampling alone, and thus this does not represent a large additional expense.

When additional samples are later taken from the same site at the same sampling locations sampling alone may be acceptable. An open field with an unobstructed view of the surroundings and no indication that the field was heavily contaminated might be such a field. Even under these conditions, however, it is a good idea to have ready communications available.

4.6.2. Goggles

Almost all surfaces of the body need to be covered during field sampling, and this is particularly true of the eyes. Goggles that protect all around are preferred over visitor-type glasses (See Figure 4.3.) Protective eyewear is especially needed in areas that are heavily contaminated or are suspected of being heavily contaminated. They are also needed when working with drilling or powered sampling equipment.

Workers have many reasons and excuses as to why they should not have to wear safety glasses or goggles. Often the complaint of workers is that they cannot see through them. After wearing them for a couple of hours, however, they often forget to take them off when they leave the work

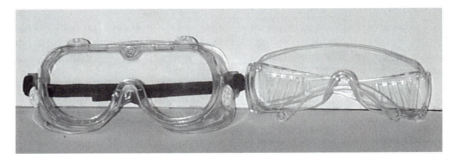

FIGURE 4.3 Safety goggles (left) and visitor-type safety glasses (right).

area. Another argument is that the person will not spill anything and so there is no reason to wear eye protection. In this case the counterargument is that they cannot say the same thing about the people with whom they are working.

Goggles in the Laboratory

During the past 30 + years I have taught students laboratory skills and safety. In that time I have never had a student lose his or her eyesight. For the whole 30 + years students constantly complain about safety goggles. The most common complaint is that they cannot see with them on. After several laboratory sessions, however, students often forget they are wearing goggles and forget to take them off when exiting the laboratory!

Reminding workers and students to wear goggles is a constant chore. Even after 3 years of reminders I will walk into a laboratory and find students not wearing their goggles. I never allow anyone to work or even be in the laboratory without goggles.

Workers have two valid complaints about wearing goggles. One is that in hot weather they fog up. The second is that they leave marks on their faces. Fogging is a problem in hot, humid conditions. Antifogging goggles are available, and workers can leave the work area and remove the goggles and allow them to air out if need be. If marks are left on the face then the straps are not adjusted correctly. Goggles need only be tight enough to prevent liquids and solids from getting into the eyes; they generally do not need to be tight enough to leave marks.

Protective glasses or goggles must have shatterproof lenses. They must protect the eyes not only from the front but also from the sides, bottom, and top. It is surprising how often liquids and solids can splash and bounce around in such a way that they enter the area of the eyes from the side, top, or bottom. Workers should be encouraged to maintain their goggles, which means keeping them clean and sanitary at all times. Dirty lenses do obstruct vision, and dirty sides may induce irritation and dermatitis. Cleanliness is a must for the productive use of safety goggles.

4.6.3. Face Masks and Respirators

Whether in enclosed areas or open fields, very dusty situations call for the use of simple filtering dust masks. These are made of paper and cover the nose and mouth. Dust masks are designed to simply filter dust particles out of the breathing air; they must not be used in situations in which gases or fumes are present because they will not protect personnel from these dangers.

In some situations the nose and respiration system must be protected from toxic, overpowering, or otherwise hazardous gases. In these cases a respirator, which can come in two forms, is required. One covers just the nose and mouth and is called a half mask respirator. (See Figure 4.4.) The other covers both eyes, the nose, and the mouth and is called a full face mask. Both types come with a variety of filters suitable for different situations. There are filters for organic vapors, acid vapors, ammonia and amines, dust, and various combinations of these. In all cases it is essential that they be adjusted to fit properly. They must also be checked for leaks before they can be effective in protecting the wearer.

In agricultural sampling respirators should always be readily available. Anhydrous ammonia, which is a gas at standard temperature and pressure

FIGURE 4.4 Half mask respirator: A—absorbing cartridge/filter; and B—body of respirator.

(STP),* is used as a fertilizer. It is stored, transported, and applied in a liquid form from pressurized tanks. A leak from any pressurized tank is a potentially lethal situation, and anhydrous ammonia is no exception. Ammonia fumes are choking. They cause eyes to water, and can quickly overcome workers. A complete face-covering gas mask designed for ammonia must be available when handing ammonia and must be worn when contending with ammonia leaks

Gaseous ammonia reacts with water in air and soil to form ammonium hydroxide, which is a strong base. Highly concentrated ammonium solutions continuously give off ammonia gas. Because of these characteristics it is possible to find pockets of ammonia-saturated water in soil after a spill. Disturbing these pockets can release enough ammonia gas to overcome a person sampling the field. Many other situations that may release noxious gases can be encountered in field sampling, and this makes having a respirator readily available essential for safety in field sampling.

4.6.4. Gloves

There are two general types of gloves, those designed to protect from heat and abrasion and those designed to protect from chemicals. When handling chemicals or environmental samples chemical protective gloves are used. These are either the familiar surgeons' rubber gloves or a synthetic rubber glove often reputed to be more resistant to chemicals. These gloves are not abrasion-resistant, and care needs to be taken to prevent tears and holes, which render them totally ineffective. A second type of rubber glove is the household type. This glove is more abrasion-resistant and often bulkier, making it hard to handle equipment. It is not intended to be used to protect the wearer from chemicals.

There are several important things to remember about chemically protective gloves. Some chemicals are known to penetrate both the gloves and the skin, and enter the bloodstream. For instance, it is known that amines and methyl mercury will penetrate rubber gloves, penetrate the skin, and enter the bloodstream. No type of rubber or synthetic glove can claim to give 100% protection from all chemicals. Indeed, it is difficult if not impossible to determine which chemicals can pass through a glove and onto or into the skin. Using gloves with a sense of impunity is dangerous, and hands should always be washed thoroughly after handling chemicals or contaminated samples, regardless of the type of gloves used [10].

---

*Standard temperature and pressure is 0°C (32°F) and 101 kPa (1 atmosphere).

4.6.5. Other Protective Clothing

In all sampling situations one must wear clothing that covers the body. As the severity of the pollution increases, however, the protectiveness of the clothing must increase. Long pants, a work shirt, and socks and boots are sufficient for agricultural field sampling for plant nutrients. For a highly contaminated industrial site a "moon" suit with self-contained breathing filters or tank of air is a must. The material of the suit must be resistant to the types of chemicals, acids, bases, solvents, and volatile organic carbons (VOCs) likely to be encountered in the contaminated field [11].

4.6.6. Decontamination

In agricultural situations, including soil sampling, clothing can be cleaned by normal means. In fields contaminated with herbicides, pesticides, and other hazardous chemicals, however, arrangements must be made for decontaminating persons, clothing, and equipment at the field location. It is not responsible to allow workers to wear or carry contaminated soil, water, clothing, or tools out of the field [9].

4.6.7. Bleeding

In many societies bleeding is considered bad and must be stopped immediately. In any situation in which chemical or biological entry into the wound is likely, bleeding is good because it washes out the affected area. In this case we are talking about a few milliters of blood, not liters. The wound must also be thoroughly washed with soap and water, however. This is similar to the prudent use comments below. Once washed a protective covering can be applied.

A first aid kit is an essential piece of safety equipment. While safety kits contain a large variety of high-quality bandages and other medical supplies, I would highly recommend making sure that they contain at least one liquid bandage kit. This is a Superglue-type material that is spread on a wound. It holds the sides of the cut together, seals the wound, and sterilizes it—all without pain. This type of bandage can keep cuts stabilized and in good condition until proper medical help can be arranged.

4.6.8. No Food or Drink

No food or drink should ever be allowed in the field to be sampled, in storage areas, or in the laboratory. This is to make sure that no one ingests an injurious component. It is all too easy to lay a sandwich on a contaminated counter and then pick it up and eat it or to drink out of a container thinking it contains water only to find that it contains or previously contained arsenic solutions. Do not allow even unopened food or drinks in any of these areas. Also, do not allow empty food containers to be brought into these areas. Seeing a food container in an area is tacit approval for eating in that area.

In addition to preventing contamination of food with the sample, contamination of the sample with food is prevented. Food contamination of field samples can cause a lot of confusion in the laboratory and in the interpretation of laboratory analytical results.

4.6.9. Toxic Fumes

Toxic fumes can be encountered in any situation. Most often, however, of primary concern is a person being caught in a confined space and overcome by fumes. This could happen in the bottom of a soil pit as well as a closed room. Even if the top is open, any confined space is of concern. It is also possible to become exposed to toxic fumes when opening a well casing or by hitting a pocket of gas trapped in soil. All wells should be suspect in terms of containing toxic gases.

The first thing to do if a person is exposed to any fumes is to remove the person to fresh air. This means outside the building or enclosure and away from the source of the fumes. If the person continues to show signs of exposure, he or she should be taken to a hospital.

4.6.10. Prudent Use

All chemicals (and some things not thought of as being chemicals, such as water) can be dangerous if used inappropriately. A person can die from drinking too much water. Also, many people drown each year. Observing prudent use and proper safety precautions—not swimming alone, for instance—should always be followed.

Prudent Use of Chemicals in the Laboratory

At the beginning of a laboratory exercise I was telling my students that the chemicals we would be using that day were not particularly toxic or dangerous. A student spoke up and said that if she drank a beaker of one of the chemicals she would die. I answered that that was true but that my comments were intended to be understood in relationship to using the chemicals in the manner and amounts appropriate to the laboratory exercise they were to be doing and that in any case drinking in the laboratory is prohibited!

4.7. MONITORING

Monitoring will be an important component in any field sampling plan. Monitoring will be of both the environment and the personnel doing the sampling. Methods of monitoring will be different for each situation.

4.7.1. Environmental Monitoring

Environmental monitoring will involve monitoring both air and water. The air in and around the field will need to monitored for dust and escaping gases. In many cases these monitors will not need to be elaborate and may be electronic gas and particle detectors incorporated into a field weather station. In addition to air sampling water sampling may also be needed. This will entail installing one or several monitoring wells and sampling the water frequently. It may also involve sampling surface runoff water if there is significant rainfall in the area or the area is irrigated.

4.7.2. Personnel Monitoring

Personnel may need sampling devices on their person to measure their exposure to contaminated materials. This may be as simple as making sure that persons working in a field do not get contaminated materials on their skin. It may also mean making sure they are not exposed to harmful vapors or gases.

If exposure does occur the affected part of the body is washed thoroughly with soap and water. Thoroughly means that water is allowed to wash over the affected area for at least 5 min. If massive exposure occurs all clothes must be removed from the person—without regard for modesty— and the person washed in a shower. The clothes then need to be disposed of properly or washed thoroughly.

Portable gas sensors are readily available and vary from a simple chemical test to sophisticated electronic sensors. Indicating badges or tubes filled with indicating materials are easy to use. Some indicators require air to be pulled or pushed through them, while others are passive. There are also several types of electronic gas sensors available. If gas sensors are deemed necessary, then gas mask or respirators, either half or full face, must also be readily available at all times. Gas masks and respirators come with a variety of filters designed for various types of gases or vapors, thus the type of filter chosen must match the types of vapors likely to be encountered.

4.8. SAFETY AT DIFFERENT FIELD TYPES

There are four types of fields that may need to be sampled. One is the agricultural field, in which soil or feed is sampled, to determine the level of plant or animal nutrients or various agricultural chemicals present. The second type is the abandoned industrial or waste field. Such fields are likely to contain a mix of dangerous or toxic chemicals. The third is an active industrial field in which a spill or accident has occurred or is occurring. The fourth situation is where there has been a spill or contamination of water. In all the situations described below a CHP is essential. Such a plan (see below) involves developing procedures for the safe handling of chemicals and chemical emergencies [3,12,13].

4.8.1. The Agricultural Field

An important safety precaution in sampling agricultural sites is to remember that they are usually remote from both medical facilities and rescue services. Treatment for any type of injury—physical (e.g., falling down and breaking a leg) or chemical (e.g., being enveloped in a cloud of anhydrous ammonia) thus may be hours away. If you are only 30 min from the nearest rescue squad, it will take them 30 min to get to you and another 30 min to get you to the hospital!

Agricultural sites are often fields being sampled primarily for plant nutrients. Although such fields are relatively safe (for exceptions see below), there are still some safety precautions that must be taken. The first is that you should not sample alone. Even on a farm on which all the surface of the land can be seen it is not safe to sample alone, particularly if it is the first time sampling the field.

It is usually easy to determine what types of chemicals one is likely to encounter on a farm. Indeed, the farmer should have MSDS sheets on the fertilizers, insecticides, herbicides, and so on that he or she possesses and uses. In this situation a limited number of MSDS sheets is needed. Two words of caution are needed here. There are a tremendous number of old barns situated on agricultural lands. Many of these were originally associated with houses that are no longer present. These old barns can contain all manner of dangerous materials. Unless instructed to sample in and around an isolated barn, stay away from them. (Also, see the section below about industrial-type situations on farms.)

Proper clothing, which is capable of protecting the wearer from stickers and brush, must be worn even if the weather is hot. Good-quality socks and sturdy boots are also required. Other than that, sunscreen and insect repellent can be used as required by location and conditions. It is often handy to have a jacket or vest with numerous large pockets for carrying samples.

In some cases an agricultural sampling situation may involve a farm with animals (horses, pigs, cattle, goats, sheep) as part of the farming operation. Although farm animals are domesticated they are not safe to be around, especially if one is not familiar with handling them. All of these animals are capable of killing, maiming, and in the case of pigs eating humans. Sampling should never be attempted in fields containing animals without the farmer's permission and preferably with the farmer present. These same precautions can be extended to areas in which the person doing the sampling is not familiar with the animals likely to be present. In this case (e.g., remote wilderness areas, game preserve parks in Africa), a knowledgeable guide should accompany the sampler at all times.

Agriculture fields can quickly and seamlessly develop into industrial-type sites. An unusually high level of nitrogen in a soil sample may be traced back to a leaking tank of liquid fertilizer. At this point a field sample needed to access plant nutrient levels has led to an industrial-scale problem of sampling around a large tank of potentially dangerous chemicals to find the leak.

4.8.2. Abandoned Industrial or Waste Field

As noted above, this might be an abandoned industrial field or a field in which chemical or other wastes have been discarded, or it may be an active manufacturing plant. Such fields have similar safety concerns. One of the most important is that large quantities of concentrated toxic chemicals are likely to be present. These may be above ground and easy to see. It is equally likely that they are in below-ground storage tanks, however. The condition of the tanks will probably be unknown and they may or may not be leaking. Samples from deep in the soil and from monitoring wells thus may be needed.

Toxic chemicals can have a wide range of dangers associated with them, depending on their characteristics. It thus is not good enough to say that a field contains toxic chemicals; one must know what type of chemicals are involved. For this reason it is prudent to consult reference information about the types of chemicals likely to be present. Such information can be found in MSDS sheets and the *Merck Index* described below [6]. The situation here is very different from that described above for agriculture because the full range of possible compounds will not be known.

In this case a large number of chemicals may be involved. For that reason a large number of MSDS sheets need to be readily available. Also in this situation there are likely to be surprises; chemicals, tanks, barrels, and so on not previously reported as being present may be found. From a safety standpoint it is important to be prepared for any eventuality.

4.8.3. Active Industrial or Manufacturing Field

Sampling in and around active industrial or manufacturing sites should be similar to agricultural sites in that you should be able to obtain an accurate list of the chemicals present. The difference is that there will be large concentrations of highly dangerous chemicals. The proper safety clothing for such a site should be known and must be strictly adhered to.

4.8.4. Contaminated Water

Safety precautions discussed above are generally applicable to contaminated water. If sampling water involves using a boat on a lake, river, or other body of water, however, all water safety precautions, such as life preservers, must also be followed.

4.9. CONTAMINATION TYPES AND CONCENTRATIONS

Two other aspects of safety are the contaminant concentration and type. We are all exposed to radiation and arsenic every day but we do not feel it and there are not any ill effects from them as long as they are at natural or background levels. There are some toxic chemicals that are an essential part of the diet. Selenium is toxic and yet it is essential for good health. Boron is an essential nutrient for plants and yet if levels are even a little too high it is toxic to them. (See also Chapter 9.)

When concentrations above natural or background levels are likely to be encountered in the field safety precautions are extremely important. On a site heavily contaminated with VOCs it is critically important to eliminate all sources of ignition, such as matches and cigarettes. In fields contaminated with heavy metals or selenium, inhalation of dust is of concern. Fields contaminated with radioactive waste require special protective clothing and radiation monitoring badges. In addition to knowing the type of toxic material likely to be encountered it is thus also important to know its likely concentration.

Many types of contamination are designated toxic, but that does not mean that they are all toxic in the same way. Gases are toxic by inhalation; liquids by inhaling fumes, drinking, or getting on skin; and solids by contact, eating, or reaction to form toxic substances. Because of these differences the way they are treated and the precautions taken need to be appropriate to the material being handled. It is thus important to consult MSDS sheets dealing with the particular toxic material present or being handled.

4.10. OFF-FIELD PROTECTION

Sample containers need to be clean on the outside before they exit a contaminated site. If glass containers are used they should be sealed in a plastic bag and wrapped in protective material, such as bubble wrap. No portion of the sample should escape during storage or transportation to the analytical laboratory. Also, as noted above, the proper signage needs to be affixed to containers of samples so that in the case of a spill persons handling the samples will know how to respond appropriately.

4.11. THE CHEMICAL HYGIENE PLAN

The CHP will be a document that outlines appropriate safety procedures for the field being sampled. It explains how to handle various types of chemicals and what to do in case of spills and accidental exposure to dangerous materials. All personnel need to know where the CHP is and how to use it. Some topics commonly covered in a CHP are given in Table 4.6. Details of the development of a CHP can be found on the Internet, along with a sample plan. There are also various publications describing CHP plans [3,12,13].

TABLE 4.6   Typical Contents in a Chemical Hygiene Plan

Recognizing hazards
Handling laboratory chemicals
Control measure implementation
Protective equipment for personnel
Training
Laboratory operations approval
Handling extremely dangerous materials
Spill response
Reporting accidents
Medical assessment
Record keeping
Appendices

4.12. ENVIRONMENTAL PROTECTION

Environmental protection can be thought of as protecting the people, animals, plants, and general area around the field to be sampled. This must include taking measures to prevent contaminated material from exiting the field (e.g., caution must be exercised to make sure that contaminated dust brought into the air by sampling does not exit the site). Likewise, sampling procedures must not lead to situations in which contaminated material can be eroded off the field by water and thus contaminate surrounding areas.

In addition, samples to be taken off the site for further analysis must be packaged appropriately (i.e., in such a way that material cannot leak out during handling and transport). (Also see clean up upon exiting above.) This is important for the safety of the people handling the samples, but also for the integrity and validity of the samples.

4.13. MSDS SHEETS

Material safety data sheets are designed for industrial situations in which large amounts of highly concentrated material are present at one time and in one place. As can be seen in Table 4.7, the information provided is complete and thorough. Not all the information may be applicable to all sampling

TABLE 4.7   MSDS Sheet Contents

Section	Topics covered
1	Chemical product and company identification
2	Composition and information on ingredients
3	Emergency overview
4	First aid measures
5	Fire-fighting measures
6	Accidental release measures
7	Handling and storage
8	Exposure controls, personal protection
9	Physical and chemical properties
10	Stability and reactivity
11	Toxicological information
12	Ecological information
13	Disposal considerations
14	Transport information
15	Regulatory information
16	Additional information

situations. An MSDS sheet discusses the fact that zinc may explode or ignite on contact with water. This is not applicable to zinc ions in soil that are at very low concentration, however. Material safety data sheets must be interpreted in relationship concentrations and form of material present and to the activity undertaken [5].

4.14. *MERCK INDEX*

The *Merck Index* is a particularly valuable resource because it provides information about the chemical properties, uses, and toxicity of a chemical. For instance, arsenic is toxic, but it is also commonly used in many items we come in contact with everyday. Additional important information is the LD_{50} of chemicals. The LD_{50} is that amount (dose) of a chemical or element that when administered to 100 animals will cause fifty of them to die. Also given is the method of administering the dosage. It may be orally, as in drinking, inhalation, as in breathing, dermal adsorption, or by injection. This gives an indication of the part or parts of the body that must be protected.

Dosage gives one, and only one, indication of how dangerous a compound is. Dose is given in kilogram, gram, or milligram (kg, g, mg) of compound or element administered per kg of animal body weight to obtain the LD_{50}. If a compound has an LD_{50} of 1 kg per kg of body weight it is *probably not, but not certainly*, as dangerous as a compound that has an LD_{50} of 1 g per kg of body weight. This compound in turn is probably not as dangerous as a compound that has an LD_{50} of 1 mg per kg of body weight. The word probably is used because some chemicals may have long-term or other detrimental effects not assessed by the LD_{50}.

The method of administering the dose is also of importance. Inhalation is perhaps one of the most dangerous and insidious ways materials enter the body. Gases, which are often undetectable, can be breathed and toxic symptoms or death result. Ingestion and injection both require that the person actively intake the material. This can but should not happen, but accidents occur. As noted above one way to prevent this is to make sure that people do not eat or drink where chemicals are being used or stored or are in the process of being transported. This includes samples, which must not be stored near food or in food containers [6].

4.15. CONCLUSIONS

Safety is extremely important in field sampling. Development of safety procedures and a chemical hygiene plan are needed. Personnel protection, including safety glasses, proper clothing and gloves, and other safety equipment must be available to all. Safety training is a must, as is continued emphasis on safety. Safety training should be a continuing part of the safety plan. With proper safety preparation and execution a minimum of safety problems will occur, and this will result in savings of time and money.

QUESTIONS

1. List and describe two major types of field sampling.
2. List and describe the general safety rules to be followed at all times.
3. A person is found dead next to a container that apparently contained liquid nitrogen. Explain how this could have happened considering the fact that nitrogen is not toxic. (Hint: A mole of any gas occupies a volume of 22.4 liters.)
4. Explain why, giving some examples, it is important that at least two people be present during sampling.
5. Fumes overcome a person. What is the first aid step to be taken?
6. What in general are the chief differences between sampling an agricultural field and an industrial field?
7. Why is it that gloves do not always protect?
8. Using a chemical supply catalog look up personal gas monitoring devices and describe two that function on different principles.
9. Describe STP.
10. Describe in detail the contents of a CHP.
11. Explain tools used in off-field monitoring.
12. A compound is listed in the *Merck Index* as having an LD_{50} of $10\,g/kg$. What does LD_{50} mean?
13. What do the following stand for?
    (a) MSDS
    (b) VOC
14. Give some examples of data important to sampling that are given in an MSDS sheet.

REFERENCES

1. Environmental, Health and Safety. http://www.osha.com.
2. National Safety Council Online Resource/Library. http://www.nsc.org/library.htm.
3. National Research Council Recommendations Concerning Chemical Hygiene in Laboratories (Non-Mandatory)-1910.1450. http://www.osha.gov.
4. Environmental Management. http://www.epa.gov.

5. The Physical and Theoretical Chemistry Laboratory. Oxford University. Chemical and Other Safety Information. http://physchem.ox.ac.uk/MSDS.

6. The Merck Index: An Encyclopedia of Chemicals, Drugs, and Biologicals. 13th ed. O'Neil MJ, Smith A, Heckelman PE, Obenchain JR Jr, Gallipeau JAR, D'Arecca MA, eds. Whitehouse Station, NJ: Merck and Co. 2002.

7. First Aid and CPR. National Safety Council. 3rd ed. Boston: Jones and Bartlett, 1996.

8. Reynolds EM, Randle O. Pocket Guide to Safety Essentials. Washington, DC: National Safety Council, 2002.

9. Henry TV. Decontamination for Hazardous Materials Emergencies. Albany, NY: Delmar, 1998.

10. Forsberg K, Mansdorf SZ. Quick Selection Guide to Chemical Protective Clothing. 3rd ed. New York: Wiley, 1997.

11. Martin WF, Gochfeld M. Protecting Personnel at Hazardous Waste Sites. Burlington, MA: Butterworth-Heinemann, 2000.

12. Developing a Chemical Hygiene Plan. JA, Young WK, Kingsley GH, Wahl eds. Washington, DC: American Chemical Society, 1990.

13. Works R, Morrison J. Chemical Hygiene Plan for Chemistry Laboratories. Department of Chemistry Laboratory Safety. Illinois State University Office of Environmental Health and Safety. http://www.che.ilstu.edu/ChemSafety/CHP.htm. 1995.

5

Sampling

When sampling, local rules apply.
No single, simple set of guidelines is applicable to all sampling situations.
General sampling principles will need to be adapted for each field and
  sampling situation.

When sampling, it is important to first keep general sampling basics in mind. The most important of these is that the environment changes over distance and time. The second is that information about a field may be obtained either without (noninvasive) or with (invasive) sampling; in either situation, it is essential to know beforehand from where the samples are to be taken. This information is essential because it is used when relating the sample to the whole environment. It is also critical if we wish or need to sample this same field again at a later date or to remediate it. All data must be readily available, and thus it is essential to have a place to put field data, such as a map of the area, a project notebook, a computer, or all three.

Data about a field can be obtained by both noninvasive and invasive sampling. Noninvasive sampling is carried out using ground-penetrating radar (GPR), remote sensing, or other noninvasive technologies. Invasive sampling is accomplished using hand- or machine-driven samplers that extract a portion of the medium about which we wish to obtain information.

The sample thus obtained is subjected to laboratory analysis. Both methods are valuable and sometimes essential in field sampling. Both have limitations, and the best sampling results are obtained when all ways of investigating the medium of concern are used. In all cases knowing the location of sampling activities is essential. Determining the location is best done using a global positioning system (GPS). Because of its importance, GPS will be discussed before examining either noninvasive or invasive sampling.

In many cases noninvasive sampling is done before invasive or physical samples are taken, mainly because these methods show surface and subsurface features affecting the invasive sampling pattern. In some cases these methods are necessary for the safety of sampling personnel and buried utilities. They will thus be used to develop the invasive sampling plan. Because they will be used first they will be discussed before discussing actual invasive sampling.

Data obtained by all sampling methods can be combined in a geographical information system (GIS). Combining data allows them to be both stored and displayed on maps of the area of interest (e.g., thematic maps). As many or as few data as needed can be displayed at one time. Additionally, a GIS system can relate data obtained from sampling to data obtained from other sources, such as aerial maps and soil surveys.

5.1. GENERAL SAMPLING CONSIDERATIONS

Analytical and instrumental methods of analysis are very precise and accurate, but the results of analyses of environmental samples are not. This means that the variability comes either from the sampling procedures or during sample handling and storage prior to analysis. Sampling is the most important source of variability, and both the sampling and handling process must be carried out with great attention to detail.

Sampling is the act of isolating a portion of a larger entity, analyzing it, and using the analytical results to describe the characteristics of the whole entity. In statistical terms the sample is an individual, which is a member of a family of similar individuals. Analysis shows that even environmental samples taken from sites close together may not be individuals or members of a family. When samples are taken at progressively increasing distances from a specific location they change in one of several ways. Characteristics can change gradually until a very different individual is identified, or they may change in a repetitive manner over long distances.

On the other hand, samples may show gradual or repetitive changes and all of a sudden show a dramatic change in a completely different direction. For example, a soil at the top of a hill might be well developed, with distinct A, E, and B horizons. As expected, the B horizon is high in clay, but there is good internal drainage. The hillside profile might have similar horizons except that they would be thinner and less well developed. At the bottom of the hill next to a stream the soil has a thick, dark, high organic matter A horizon underlain by a sandy B horizon with a small amount of clay in it. There is thus a gradual change down the hill and an abrupt change at the bottom.

It is essential that such changes in the environment be kept in mind both when designing a sampling plan and during noninvasive and invasive sampling. If unexpected changes occur, they are recorded in the project notebook and accounted for in the sampling procedures, sample analysis, and the interpretation of sample analysis.

In sampling, air and water variability is not as great as with soil. The atmosphere is particularly uniform, but the hydrosphere is more variable in that the salt content of water varies with location. It is important to remember, however, that there are layers and definite direction in movement in both of these spheres. Downwind or downstream from a source of contamination it is dramatically diluted, and thus changes in location, either horizontally or vertically, may lead to sampling different layers, and this may result in differing and perhaps confusing analytical results. These sources of possible variation and error must be kept in mind when designing sampling plans for air and water as well as soil [1].

In the United States sampling of hazardous fields and sites falls under the purview of the United Stated Environmental Protection Agency (USEPA). This organization has a publication, *Test Methods for Evaluating Solid Waste Physical/Chemical Methods* (USEPA SW-846 3rd ed., November 1990). In addition, *Standard Methods for the Examination of Water and Wastewater*, published by the American Public Health Association, American Water Works Association, and Water Environment Federation (APHA, AWWA, WEF; 20th ed., 1998 and 1991 supplement) can also be consulted. Numerous methods for sampling and assessment are available from the American Society for Testing Materials (ASTM). (See Appendix B.) For health and safety issues related to sampling, the Occupational Health and Safety Administration (OSHA), particularly OSHA–29 CFR 1910.120, should be examined. Also in the United States, each state has its own environmental protection agency (EPA), which will have state-specific regulations for such things as sampling [2–4]. Most countries will have similar organizations, and these must be consulted.

5.2. GLOBAL POSITIONING SYSTEM

It is highly recommended that GPS be used in locating sample sites. As discussed in Chapter 3, the first step in sampling is to set up the coordinates for the field. This means first deciding which coordinate system to use and finding the coordinates at the corners and edges of the field. These are then put on the maps of the fields, which are then used for all other sampling activities. Maps such as those shown in Figures 5.1 to 5.3 should include these coordinate references at each corner as shown, because these locate the field unequivocally.

A GPS works using satellites in orbit around the Earth. These satellites transmit signals that are received by GPS units on the Earth. Using a minimum of two signals, a GPS unit can show where it is located on the Earth's surface. (It is highly recommended, however, that a unit accessing twelve or more satellites be used.) This system can also give the altitude, and if it is moving, the direction and speed of movement. The GPS location given is a map location, and so this system is directly linked to maps of the Earth's surface. The maps must have the same units of geographical reference as displayed by the GPS unit for both to be used together. Such

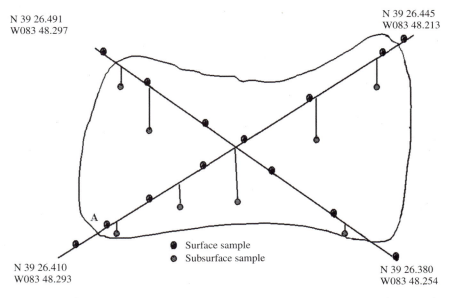

N 39 26.491
W083 48.297

N 39 26.445
W083 48.213

A

● Surface sample
◉ Subsurface sample

N 39 26.410
W083 48.293

N 39 26.380
W083 48.254

FIGURE 5.1 Transect sampling. The curved line encloses the suspected contaminated area. Note that the samples along the transect lines are taken randomly.

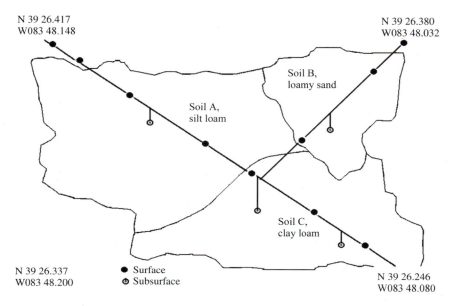

N 39 26.417
W083 48.148

N 39 26.380
W083 48.032

Soil B,
loamy sand

Soil A,
silt loam

N 39 26.337
W083 48.200

● Surface
◉ Subsurface

Soil C,
clay loam

N 39 26.246
W083 48.080

FIGURE 5.2 Transect sampling when two or more distinct soil types are present. Note that both surface and subsurface (dots) samples in all three soil types are taken.

units can also show the direction one needs to go to get to an identified place on a map or on the Earth's surface.

5.2.1. Coordinates and Time

Global positioning systems depend first on a datum, second on such coordinates as longitude and latitude, and third on time. There are different data for different parts of the world, but in North America, the 1927 North American datum is used (commonly referred to as NAD-27). (There is also an NAD-83.) In addition to the usual longitude and latitude used on maps and globes, a coordinate system called the universal transverse mercator (UTM) is also in use and must be available in the GPS unit. Before starting to use a handheld GPS unit, make sure that it is reading the correct datum and is giving locations in the appropriate format for the maps being used.

Time is kept using atomic clocks in satellites, and, in turn, time is used for determining the position of the GPS receiving unit (i.e., GPS time). Using this very precise time satellites can be kept synchronized and GPS units convert GPS time to the universal time coordinate (UTC), which is the same as Greenwich mean time (GMT). From this point, the local time can

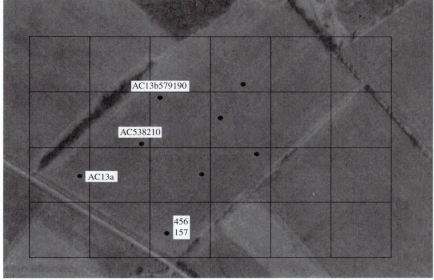

N 39 26.700
W083 48.400

N 39 26.700
W083 48.000

AC13b579190

AC538210

AC13a

456
157

N 39 26.200
W083 48.000

N 39 26.200
W083 48.000

FIGURE 5.3 Sampling grid placed on an aerial photograph showing sample locations and numbers.

be obtained. A detailed description and discussion of the whole field of navigation, data, UTM, and UTC will not be covered here; however, an excellent resource covering these topics is given in Ref. [2].

Global positioning system instruments range from inexpensive handheld units with limited capacities to stationary units with multiple capabilities. The two most common are handheld units and mobile units intended to be carried on a vehicle. Antennas are an important consideration in choosing either type of GPS unit. Several designs are available, and the one chosen should be suitable to the terrain in which it is to be most often used. Also, because basic GPS is not as accurate as one might like, it is essential to have a unit that is capable of differential GPS (DGPS) or of using the wide area augmentation system (WAAS). A discussion of these methods is given in Chapter 2.

Global positioning system units are powered by a number of different sources, the most common of which are batteries, including adapters for using the battery of a vehicle. The batteries in handheld units can be either rechargeable or of the dry-cell type. Dry-cell batteries can last up to 20 hr and many units have battery-saving features. One set of batteries can thus last for a day or two of sampling. If sampling will take longer, rechargeable batteries are preferred. If the person sampling is riding a vehicle with a 12-V battery, a battery connector is preferred.

5.2.2. Sources of Error

Global positioning system units are capable of operation in overcast and other less than ideal weather conditions. There are, however, some situations in which errors can occur. These might cause only a slight error in location or they may cause errors of 1 or more km. For instance, using the wrong datum leads to a wrong position being shown by the GPS unit. Also, in coordinating the position shown on a GPS unit with a map be certain that the coordinate system being used in both cases is the same.

Reflected multipath errors occur when the signal from one satellite is received both directly and indirectly after it has bounced off an obstruction. Such errors occur where there are large mountains or buildings near the field in which the GPS is being used. If both the direct and reflected signals are used, the calculated position will be wrong. Some GPS units can automatically correct for multipath signals, but others cannot. Two precautions thus need to be taken. First, observe the surroundings for features that might reflect a signal, and second, determine if the GPS receiving unit being used can correct for multipath signals.

In actual use, handheld GPS units can be of great value. For instance, inexperienced personnel can sample large fields two ways. First, the handheld unit can be preprogrammed with waypoints. When the person doing the sampling reaches the preselected site, the unit will beep or otherwise notify him or her. The sample can be taken and the unit will indicate how to get to the next waypoint. A second way is to provide the person with a map of the area showing the location of the sites to be sampled. Even if the sites are not precisely located, the exact location can be identified if the longitude and latitude indicated on the GPS unit are recorded at the time of sampling. At a later time the same position can be found easily if the site location is clearly recorded.

If data are needed from between sampled sites, they can be either physically sampled or the concentration of the desired constituent estimated using techniques such as kriging, which is discussed further in Chapter 6.

5.2.3. Limitations in Field Sampling

Global positioning systems are extremely useful in field sampling, and no field sampling should be undertaken without them. Sample sites should not be located at distances shorter that the accuracy of the GPS unit, however; that is, a GPS should not be used to locate sample sites 3 meters apart if its accuracy is ± 10 meters. If this is done, two sites that are 4 meters apart may have the same longitude and latitude even though they are obviously not at the same location. If such longitude and latitude data are taken, erroneous data will be obtained, and when used these data will produce inaccurate and confusing results, particularly if used in modeling.

Availability and use of a top-quality GPS unit does not make all other navigational and site location tools obsolete. Maps, compasses, altimeters, pedometers, watches, and project notebooks are all still essential and will find many uses [5]; for example, a handheld GPS unit may not have a built-in compass. Figure 5.4 shows a handheld GPS unit and watch with compass and altimeter. The GPS unit will show the direction in which you are going once moving, but will not show the direction in which you should start moving to go to the next waypoint. A simple compass therefore can be a handy addition when field sampling. Many watches, such as the one shown in Figure 5.4, have various types of compasses built in, so these can be used in the field in place of a separate compass. A word of caution, however; calibrate the compass before starting to use it in the field.

FIGURE 5.4 Handheld GPS unit (top) and watch with compass and altimeter.

5.3. GROUND-PENETRATING RADAR

Before disturbing the soil surface, and especially before taking samples deep in the soil profile, a GPR map of the field can be taken. This is particularly important if the field is likely to contain utilities, gas, water, electric, or communication lines or if it has been used by armies, in which case it is likely to contain explosives.

Because of its common use in weather forecasting and in both civilian and military flying, radar is familiar to almost everyone. What is generally unknown is that it can also be used to investigate underground features without disturbing the soil surface. Radar works by sending out microwaves which, when reflected from an object (e.g., airplane), are received by the radar antenna and displayed on a monitor. The direction the microwaves came from and the distance of the object from the antenna can be calculated. Both sets of information are typically displayed on the monitor.

Ground-penetrating radar is similar to any other radar except that the microwaves are directed into the soil and the reflected waves are analyzed. Radar waves penetrating the soil are refracted at interfaces between layers or objects with different relative dielectric permitivity (RDP). This means that the greater the difference in the RDPs the more radiation is refracted and the stronger the signal received. The largest difference is between air and water. (See Table 5.1.) There is also an appreciable difference between soil and both air and water. Originally, GPR was used in areas in which the soil tends to be dry. Today it is used in both wet and dry conditions [6].

The RDP can be calculated using the equation $\sqrt{k} = C/V$, where k is RDP, C the speed of light, and V the velocity of radar in the material being investigated. The exact RDP cannot be accurately calculated when dealing with mixed materials, such as soil and regolith. Under these field conditions

TABLE 5.1 Relative Dielectric Permitivity (RDP) of Some Common Environmental Components

Medium	RDP
Air	1
Water	80
Dry silt	3–30
Wet silt	10–40
Soil	12

Source: Data from Ref. [6].

it is either necessary or highly desirable to determine k by measurement. Burying a metal bar at a known depth in the soil or regolith and passing the GPR instrument across the area in which the rod is buried can easily accomplish this. The locally applicable k is thus determined, and then used to calibrate the instrument for the medium where the survey is to be made. At this point the depth to which the GPR survey will be useful can be calculated, keeping in mind the limitations noted below. The equation used to find this depth is

$$A = \frac{\lambda}{4} + \frac{D}{\sqrt{k+1}}$$

where A = approximate long dimension radius of footprint. The footprint is the buried surface area illuminated by the radar signal. In the equation, λ = center frequency wavelength or radar energy, D = depth from ground surface to the reflection surface, and k = average RDP of the material being surveyed. Under the best conditions, the depth of penetration of radar in soil is 40 meters (approximately 130 ft). Normally, however, usable penetration is limited to 1 to 5 m (3 to 16 ft). Within this depth, the occurrence of pipes, power lines, buried objects, caves, differing layers of soil and regolith, and the depth to the water table can be observed.

Making a survey is done by passing a GPR instrument over an area in a grid pattern similar to the grid shown in Figure 5.3. In this case, however, the grid lines are typically 1 meter apart in both directions. The data obtained are displayed using computer programs. In this way a three-dimensional picture of subsurface features can be visualized. Computer programs can also be used to filter the raw radar data to give a clearer picture of subsurface structures.

Table 5.1 gives the RDP for a number of common environmental components, but does not give a figure for saltwater, which is an electrical conductor and thus does not have any appreciable RDP. Ground-penetrating radar thus cannot be used in saltwater or in salty soil conditions. In this case salts will include all the common fertilizer elements and soil amendments, such as lime or calcium carbonate. Also, highly conductive clays are not amenable to GPR methods. While GPR is useful in dry soils, it thus cannot be used on many dry-region soils because of their high salt content. Note that this can also include natural deposits of calcium carbonate, which are quite common in arid and semiarid soils.

A number of other common environmental conditions interfere with GPR. The radiation used in radar has a magnetic component, as does all electromagnetic radiation. Thus soils or other media that are magnetic or high in iron or other magnetic materials such as magnetite cannot be investigated using this method. Other common soil and regolith components such as cobbles will also cause anonymities in the GPR signal. Some subsurface features may be hard or impossible to detect by GPR. If the angle of incidence refracts the waves away from the receiver the object will not be seen. Also, in some instances the speed of the instrument over the surface may affect the clarity of the readings [6].

5.4. REMOTE SENSING

The term remote sensing can be applied whenever data obtained from the environment are transmitted from that location back to some central receiving station, processing center, or office. This definition might include the case in which photographs are taken from an airplane and the film physically delivered to the processing center. The most common interpretation of remote sensing, however, is data about the Earth gathered by satellites and transmitted back to Earth [7].

In all likelihood, remote sensing of the field being investigated has already been done, thus one should look at various archives of remote sensed locations to determine if the location of interest is there. These data archives can be accessed on the Web through the Distributed Active Archive Centers (DAAC) and the Earth Observatory System (EOS) data gateway (both of these are National Aeronautics and Space Administration, NASA, sites) [5,6]. Both these sites are extremely complicated and hard to navigate. It will take some time to find the information you are interested in; however, if the desired data are available, obtaining a copy of the images is relatively simple.

One potential problem with available data is that the images may not include the portion of the spectrum that is most important to the sampling program. If the needed data are not available, a request can be made to have an image taken of the area. It is relatively simple and not very expensive to have satellite pictures taken of specific areas of the Earth's surface. The resolution, which depends on the portion of the spectrum needed, is typically between 30 and 60 meters. When ordering an image, both the location of the area of interest and the portions of the spectrum to be recorded must be specified [7–9].

5.5. GEOGRAPHICAL INFORMATION SYSTEMS

When developing a sampling plan and collecting historical data about a field a question to ask is if there is a GIS for the field and the surrounding areas.

A GIS may contain important information about cadastral questions, such as easements and rights-of-way; the natural environment, such as hydrology, topography, vegetation, watersheds, and wetlands; and utilities, such as water lines. This information can be invaluable in developing and carrying out a sampling plan. It can save time and money, especially if it avoids breaking a power line or similar underground facility.

Any system that makes a map or representation of the Earth's surface can be a GIS. Today, however, it is generally understood that GIS refers to maps and other data stored digitally on media that can be read and analyzed by computer. Such systems allow the addition of any digital information to the map. This additional information can then be displayed in a variety of different ways to help researchers understand phenomena occurring on or below the Earth's surface. Such maps are said to be thematic.

In making thematic maps GIS makes use of elements from other mapping tools. The most important of these is CADD and AM/FM. CADD is computer-aided design and drafting, the map-drawing capabilities of which are included in GIS, but it does not have data correlation capabilities. For instance, when looking at a hazardous waste field, GIS can be used to determine how much of it is in a flood-prone area, while CADD cannot. AM/FM is automated mapping/facilities management, which is used extensively for mapping networks and network element characteristics. Again, data analysis is better carried out using GIS. Any information about a field that is in either of these two formats can be easily imported into a GIS system.

The thematic maps produced using GIS work a little like having the base map and making transparent overlays for it. Each overlay would contain data of interest. Although the data could thus be visually observed they could not be analyzed in relation to the base map or to other layers; the GIS system allows for this. In addition, GIS systems allow for the addition and removal of one or more layers as needed.

For sampling, the most important feature of GIS is the ability to display sampling sites and their characteristics on a map of the field being sampled. This allows the sampling sites to be coordinated with other features of the environment, such as water flow. This is important because water movement may indicate other areas that need to be sampled or watched.

Information displayed on a map can be entered manually into a computer system, in the form of tables, for example. Alternately, it can be obtained directly from any digital source of information. For instance, data obtained by remote sensing from aircraft or satellites. A familiar example of this use is the display of weather. The daily weather report incorporates this type of technology. Population densities, lightning strikes, and ocean temperatures are other common examples of uses of GIS. Recently, mapping Pacific Ocean temperatures using remote sensing has been used to predict the occurrence of La Niña or El Niño. Whenever data are entered into a GIS system from any source, their accuracy must be checked. This is especially important if they are entered manually.

Today, satellites are the most important source of information for GIS. Satellites can access various portions of the electromagnetic spectrum, allowing analysis and evaluation of conditions on the Earth's surface. (For more information on satellite sensing, see the previous section on remote sensing.) Having satellites use sensors to observe the Earth's surface and having these data digitalized allows for easy insertion into GIS systems.

Satellite sensors are very versatile, and different satellites have sensors for different portions of the electromagnetic spectrum. A well-know satellite used for remote sensing is the Landsat TM (Thematic Mapper) satellite. It has a broad range of sensors, which among other things can detect both different vegetation types and changes in them. These changes can be related to changes in a plant's environment, including environmental degradation. Airplanes equipped with sensors still play a role in remote sensing, and data retrieved in this way can also be inserted into GIS systems.

In most cases, a sampling plan will not require the entity doing the sampling to implement a GIS system. Desktop computers with GIS display and manipulation software can be inexpensively obtained, however. The needed GIS data, including maps, can thus be obtained from various sources and combined to obtain the information, relationships, and analysis needed [10]. One important source is local GIS offices. It is very common for even small cities, municipalities, and counties to have GIS systems, and both the mapping and tax offices know who keeps these systems and where they are kept, and there may even be a separate GIS office, which is often small and not easy to find. If such an office exists, however, it is worthwhile finding because GIS personnel can be a valuable resource in obtaining GIS data and in developing and executing a sampling plan.

5.6. SAMPLERS

There are many different types and constructions of samplers designed for each type of media to be sampled. One of the most important considerations in deciding on the type and construction of a sampler is the material that it is made of and whether or not it is compatible with the type of material and contaminant being sampled. Figure 5.5 shows a typical simple hand sampler, while Figure 5.6 shows a sampler mounted on a truck. Additional

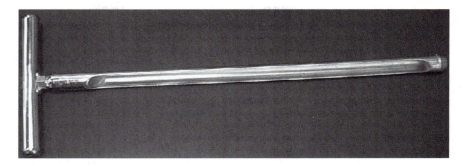

FIGURE 5.5 A simple hand core sampler for sampling up to 30 cm deep.

FIGURE 5.6 Truck equipped with power sampler. Samples range from 2.5 to 7.5 cm in diameter and up to 2.5 meters deep.

examples of samplers and information about them are given in Figure 3.6 in Chapter 3.

5.6.1. Sampler Composition

Samplers can and are made of a great many different types of materials, from plastics to metals and alloys. The choice of sampler construction material will be determined by the type of contaminant in the sample and its reactivity with sampler materials. For sampling involving organic contaminants many different materials will be acceptable as long as they do not react with the sample. When sampling is for metal contaminant the metal composition of the sampler is of primary concern. In some situations a number of different samplers of different composition will be required.

There is a great variety of plastics with different contaminant compatibilities. One widely used plastic is Teflon and Teflon-coated samplers. This plastic is resistant to almost all chemicals and has a nonstick surface. As with all plastics, however, it is not good for situations in which the sampler must cut through abrasive material. Also, if the sampler is Teflon-coated, it is imperative that care be taken to assure that there are no holes or worn spots in the Teflon, which might lead to contamination or cross-contamination of sample.

All plastic and plastic-coated samplers are best suited to air and water rather than soil sampling. Soil is abrasive and can rapidly wear away plastic coatings, which can also develop gouges and cracks, exposing the underlying material to the sample. Also, cracks can become filled with sample material and thus cross-contaminate samples.

In addition to plastic, samplers are also made of various metals, including brass, iron, steel, and stainless steel alloys. Metal samplers are common and robust; however, they can result in metal contamination of samples. This is particularly a problem when the contamination of interest is a metal ion, metal, or heavy metal. This can be even more of a problem when the material being sampled is very acid. Keep these considerations in mind when choosing a sampler and sampler material.

5.6.2. Sampler Types

There are a large number of samplers designed for different soil types and depths of sampling. Perhaps the simplest is a metal tube or core type 46 to 50 cm (18 to 20 in.) long, usually with a section cut out of one side of the tube. (See Figure 5.5.) A piece of tubing is fixed on the top, making a T, and the mouth at the bottom is shaped to make it easy to insert into soil. The sampler is inserted, usually using hand or foot pressure, into soil at the correct site and then withdrawn with the soil sample. The side opening is used for removing the sample. This sampler is a most useful in fine-textured, moist soils. When the soil and moisture conditions are favorable, it is only limited in how deep it can sample by how deeply it can be pushed into the soil using hand or foot pressure.

For deeper sampling a larger, more robust sampler of the same basic design can be inserted into soil using a manual or powered hammer. For even greater depths, a similar core-type sampler can be attached to the back of a pickup truck equipped with a hydraulic or mechanical system for insertion into soil. (See Figure 5.6.) The hydraulic press inserts a length of pipe into the soil; when removed, the core is pushed out, again using the hydraulic or mechanical system. In this way, cores of various depths can be easily obtained. This sampler can be used to quickly and easily sample large areas. (See Appendix B for sources of samplers.)

Another type of sampler is a rod about 46 to 50 cm long on the end of which is attached an auger, which is like a wood drill bit. (These are shown in Chapter 3, Figure 3.6.) These samplers are most commonly used and useful for obtaining samples in soil containing gravel or other large stone fragments, in which they can be used to depths of 15 to 20 cm (6 to 8 in.). Auger-type samplers can have the same variations as the core samplers mentioned above.

Another type of auger sampler is one with an integral bucket. These samplers are screwed into the soil and a sample is collected in the bucket (Chapter 3, Figure 3.6). The sampler may be fixed to a T-type handle with a definite length of pipe between the handle and the auger. Some people keep a series of samplers of different lengths for different jobs. In situations in which there is a constant need for samplers with different lengths the lengths are threaded in such a way that additional pipe can be added or removed to make the sampler longer or shorter.

For very deep sampling in complex situations it is imperative that a standard well-drilling outfit be used [11,12].

No matter which type of sampler is being used, it is important to keep samples separated and identified. Samples should not be allowed to become cross-contaminated during the sampling procedure. The sampler therefore needs to be cleaned after the collection of each sample.

Generally speaking there is no correlation between the precision or accuracy of analytical results and the type of sampler use. Because of this, soil scientists familiar with the area in which the field is located can give excellent guidance as to the best type of sampler to use for the local soils and regolith [1].

5.7. SAMPLE AMOUNT

The amount of sample taken will be determined by two factors. How much sample is needed for analysis, and what type of component is the sample to be analyzed for? Because of these considerations the absolute size of the sample cannot be given. The amount of sample needed in the field laboratory and the commercial laboratory are added to obtain the total amount of sample needed. This may change, however, depending on the sample container requirements, because some samples must fill the container completely and this determines the amount of sample taken. In those cases in which multiple analyses are called for, more sample is also needed.

In sampling, several "cores" will be combined to make a sample to be analyzed. For instance, three or four cores might be taken from each grid in Figure 5.7. These would be placed in the same bag as a single sample for analysis. Figure 5.8 shows three cores and a core still in the sampler. If the cores are not about the same size, the larger core sample characteristics will dominate the sample and analytical results. This can lead to dilution or concentration of the component of interest. In either case the resulting analytical data will not give a true picture of the area being sampled. It is

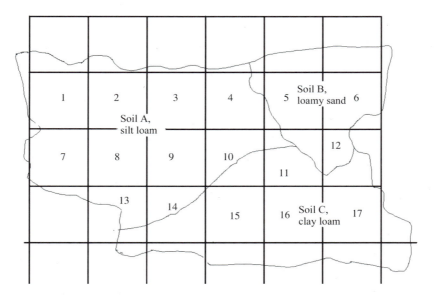

FIGURE 5.7 Grids for sampling. Both surface and subsurface samples can be taken randomly or systematically from each grid over time.

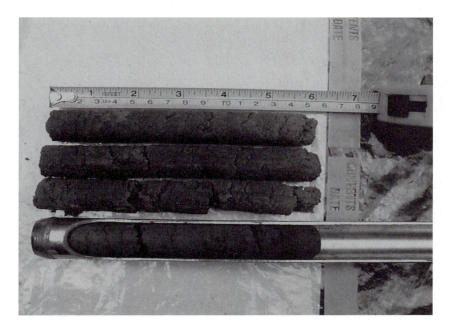

FIGURE 5.8   Soil cores with one core still in sampler.

thus necessary to used the same sampler and take the same amount of sample each time. This is true for both surface and subsurface samples.

Note that samples for volatile organic compounds (VOCs) analysis are handled very differently. Samples must be taken, handled, and contained in such as way as to minimize the loss of sample components. Some procedures call for the sample container to be filled to the top, leaving as little headspace as possible. The container is then sealed with a top that is proven to prevent leakage of VOCs. Other procedures will require more or less sample, depending on the sampling, container, and analytical methods used. The analytical laboratory can provide specifics as to sample handling, the amount of sample, and the container needed for the specific VOCs being sampled and analytical methods to be used.

When the sample arrives at the analytical laboratory it will be ground, sieved, and mixed. A subsample of this mixture is usually on the order of 1 to 10 g, although larger samples may be used in some cases. The size of the sample taken from the field thus does not need to be large. As noted above, however, all samples should be of about the same size so that observed changes are not due to changes in sample amount, but to the inherent characteristics of the sample [12,13]. Sample size can be a trap, and this will be discussed in further in Chapter 11, along with other traps.

5.8. SAMPLE CONTAINER

Almost every conceivable type and composition of container has been used to contain contaminated air, water, and soil. Glass bottles or containers are most frequently used for air and water samples, however, while paper and plastic bags are used for soil samples. Before deciding on a container material, however, it is essential to know the history of the field and have a clear idea of the types of contaminants it contains or is suspected to contain. It is also important to know as far as possible the species of the contaminant that is of primary interest. For example, when sampling soil for VOCs, bags are not suitable containers. In this case a glass container will be provided by the analytical laboratory, along with use and filling instructions. A more complete list of typical containers and their characteristics is given in Chapter 3 in Table 3.3 and Figure 3.7 [11–13].

5.9. DUPLICATE, SPLIT, AND COMPOSITE SAMPLES

Duplicate samples are taken from the same location and are used to check the accuracy and precision of analytical methods. In the case of water and air, duplicate samples may serve this purpose very well. In soil or regolith sampling, duplicate samples are two cores taken adjacent to each other. These two cores may not be identical, and indeed may be significantly different from each other. Duplicate soil samples can be different by as much as 5–10%. If the occurrence of this variation does not cause a problem, duplicate samples may be useful. If duplicate soil samples give exactly the same analytical results, the validity of the sampling and analysis is in question.

A single sample split into two subsamples placed in two different containers would represent split samples. Splitting air and water samples, which can be done in the field or in the laboratory, can be an effective tool for comparing analytical results. Soil and regolith samples must be ground, sieved, and mixed thoroughly before splitting. Thus, splitting of these samples can only be done in the laboratory. Even with this type of sample, significant variation between subsamples can sometimes be found.

Composite samples are obtained by taking two or more samples and combining them to make one sample. For instance, three equally spaced, 20-cm-deep core surface samples might be taken from grid 3 in Figure 5.7 and combined to make a composite sample for this grid. Another approach among the many variations of this concept would be to take more cores or take them randomly. The cores obtained are ground together and mixed before subsamples are taken for analysis.

The above grid may be smaller than 1 ha or as large as 5 or more ha. The idea is that this composite sample will provide an average value for the amount of contaminant or constituent in the area represented by these multiple samples. Such a sample will allow the same level of remediation to be applied to the whole area sampled as long as the contamination is fairly evenly distributed in the area in which the compounded samples are taken. This is also sometimes referred to as composting or composted samples, especially if the sample is left for some period of time before analysis [13].

When samples from some depth in a field are to be taken, the same procedures are used as described above. In this case, 15α in Figure 5.9 represents a volume of soil corresponding to the area of grid 15 and depth of α, and likewise for volume 15β. Because contaminants must leach down into the soil and because obtaining samples at some depth is significantly more difficult than taking surface samples, usually fewer samples at various depths will be taken and analyzed. In this case, the goal is to determine the depth of leaching of the contaminant, so that the volume of soil to be remediated is known. The models discussed in Chapter 7 are often valuable in deciding on the frequency and depth of sampling in these cases.

In cases in which there is a buried tank or a leaking pipe contamination will originate at some depth in the soil. In these and similar cases deep sampling will be mandatory, and a large number of such samples will be required from various depths, depending on the leaching of the contaminant.

5.10. SAMPLING STRATEGIES

There are several different strategies or ways in which people obtain samples from a field. Actual determination of specific sites will be presented in Section 5.13.

5.10.1. Transect Sampling

A first step in sampling is to obtain information about the extent of contamination. To do this, make one or several three-dimensional sampling transects of the field to be sampled. In Figure 5.1, two transects are sampled across a field that is assumed to be uniform (i.e., one soil type). In this case, surface samples are taken at regular intervals across the contaminated area. Also, samples from regular depths are taken at several sites along the transect(s). In this process it is essential to make sure that two types of controls are also taken. The first is to make sure that the first and last samples are taken from outside the area that is assumed to be contaminated. Also, samples from various depths must include a sample obtained from below the contaminated material.

In Figure 5.2 transects are taken in an area containing three soil types. Note that the transect lines are chosen in order to obtain samples from each soil type; the figure also illustrates taking both surface and subsurface samples at regular intervals. Also, some transect samples are taken outside the suspected contaminated areas. These samples are essential to establish the boundaries of the contaminated areas.

Normally it is important to take into consideration what soil type is being sampled, and this is true for the transect samples. In this case, however, it is only important to label samples so that the soil type from which the sample is taken can be determined later. This information is useful for the development of the detailed sampling plan. At this point, the most important information to gather is the level and extent of the contamination throughout the contaminated area. An advantage of transect sampling without regard to soil type is that if analytical results are consistent for the different soil types the detailed sampling plan and the final analytical work is simplified, which in turn saves resources, including money.

During transect sampling, however, unusual soil characteristics should be noted; that is, any indication that the soil types are different from those mapped or that the soil characteristics have not been noted elsewhere. If the soil is mapped as a silt loam and an area with sandy texture is encountered in the transect sampling, this is important. A sandy soil will have absorptive, retentive, and extraction properties that are very different from silt loam. Layers that are unusually hard or compacted must also be noted, because they will affect contaminant movement, analytical results, and applicable remediation procedures.

The transect sampling results can also be used to identify zones of contamination (i.e., some areas contaminated at uniform but relatively lower levels of contaminants than surrounding areas). When determining the number of samples needed, these areas can be treated separately from other areas (see Chapter 6), since they may, for example, require less sampling. This leads to increased efficiency and less cost to the project.

The analytical method chosen for analyzing the transect samples should be simple, rapid, and inexpensive. Accuracy and precision should be of secondary concern. Use of an entirely different analytical procedure for the detailed sampling and analysis is appropriate. Also, note that if the two different analytical methods give similar results, confidence in the validity of the whole sampling and analysis process is greatly increased.

Preliminary Nitrate Sampling

I worked with some researchers concerned with nitrate contamination in well water. They had ten wells they were sampling and wished to know which wells were in need of monitoring. At the time I was using a nitrate ion selective electrode (ISE) for some other work and was asked if I could analyze their samples. Some of our colleagues scoffed at the idea, saying that the nitrate ISE was not sensitive or accurate enough for this work.

In analyzing the ten well water samples, the distinction between contaminated and uncontaminated wells was obvious. The uncontaminated wells gave no response. There were three wells that gave a response of 1000 ppm nitrate (NO_3^-), however. This analysis allowed us to identify the contaminated wells and to predict which other wells should be looked at using more sensitive methods. It is important to point out that the sensitivity of the ISE that I was using was 10 ppm NO_3^-, and that the allowable level of nitrate in drinking water in some places is 10 ppm and in other places higher.

The ISE is an example of a rapid analytical method that is suitable to transect samples. Although it can be used in the field without elaborate sample preparation, the field office laboratory is a better place to make the analysis.

The ISE is similar in appearance and operation to a pH electrode. Indeed, the more sophisticated pH meters are able to accommodate many different ISEs. The use of an ISE is as quick and easy as determining pH using a pH meter. Typically the time for doing an ISE analysis is less than 1 min. Portable pH meters, which can do both pH and ISE analysis, are readily available [14,15].

Transect sample results can be combined with detailed sampling data to confirm the detailed sampling data and to make statistical calculations as to the number of samples needed. These analytical results will also provide information that can be used to determine if a sample is extraneous or is indeed part of the population being sampled. More information on this can be found in Chapter 6.

During transect sampling variations in soils and soil types in the sampling area are noted but ignored. (See Figure 5.2.) However, when the detailed sampling plan is designed, the soil and soil types present must be taken into consideration. This means that the whole sampling plan may be designed without regard to soil types. The soil types are recognized when samples are taken and analyzed, however. In this case the batch number and the number on each individual sample must be unique, so that they can be identified later if needed as being from one soil type [11].

When the transect samples have been taken and analyzed, a detailed sampling plan can be developed. This plan will take into consideration observations and analysis made during transect sampling and the statistical analysis of the analytical results. (Statistical analysis will be described in Chapter 6.)

5.10.2. Detailed Sampling

Detailed sampling can be of four types. The first is sampling to determine the levels of plant nutrients in a field. The second is sampling to determine the extent of contamination so that the contaminated material can be removed for remediation. In this case, the field will only be sampled once, and subsequent sampling in this location is not expected to occur, although monitoring may be called for. In this case, it is important to identify all the contaminated material and assure that it is safely removed. This is coupled to the third type, which involves sampling the material once it is in its new location. This will entail random sampling not related to the position of the material in the environment but to the amount of disturbance the material is subject to during the cleaning process.

The fourth detailed sampling situation occurs with bio- or phyto-remediation. In this case the field will need to be sampled several times to ascertain if remediation is progressing as planned and if additional steps need to be taken, and to certify that remediation has been completed. In this case, sampling sites will initially be randomly assigned or sited. Subsequent sampling will take place at or near these originally designated sampling sites, however. Final sampling before designating the area as clean should involve a few additional random sites to assure cleanliness.

There are several possible approaches to the detailed sampling plan. One is to make it a totally random three-dimensional sampling. A potential problem with this approach is that picking random sites to sample may have sampling occurring in many areas in which you are certain there is little or no contamination. Another possibility is that you may obtain less information about a particular field when multiple samplings are called for.

Another approach is to produce an imaginary three-dimensional grid of an area, each grid of which will be sampled. Such a grid has been superimposed on the maps shown in Figures 5.3 and 5.7. Note that in each case the position and limits of each field have been unequivocally located with the longitude and latitude of the four corners of the field.

Another approach would be only to sample within randomly selected grids in the girded area. Subsequent samples can again be taken randomly within the given grid. Another possibility would be to take individual grid elements and apply them randomly to the field. For instance, grid element

15 $\alpha\beta\gamma$ (see Figure 5.9) would be applied (with their own individual designations) to random areas in the field to designate places in which sampling is to occur [4]. It is important to remember that the grid is three-dimensional and that every sampling event must include samples from all three dimensions. (See Figure 5.9.) The fourth dimension will be determined by sampling at different times.

It is often the case that the soil and underlying regolith is composed of several layers of differing composition. These layers have different densities and may be at different depths across the contaminated site. Such layers may need to be sampled specifically, and the samples taken will need to contain the same amount of contaminated material. This will necessitate sampling at different depths at different points across the contaminated field. It will also necessitate taking a larger sample in some cases (low bulk density), or smaller samples in other areas and at other depths (high bulk density strata; see Chapter 2) [11].

5.11. TOPOGRAPHY

A totally random sampling plan has the advantage of removing bias from the sampling. In many situations, however, the topography of the landscape (see Chapter 2), including that of the bottoms of bodies of water, affects the level, depth, location, and concentration of the component of interest. For this reason, it is important that topography be taken into consideration when the sampling plan and sampling sites are chosen. A sloping field

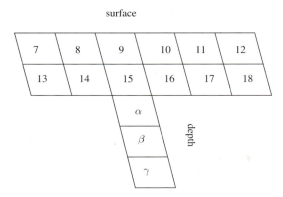

FIGURE 5.9 Numbering across the top for surface samples. Greek letters designate subsurface samples.

should be sampled differently from a flat field. A river bottom sampling plan should be different from sampling a plateau above a river.

In the case of a sloping field, the top, middle, and bottom of the field should be designated as separate sampling areas. If grids are used, a horizontal grid line should occur at the top and bottom of the slope, with additional grid lines in the middle as needed. A field with little slope can be sampled without regard to slope. However, such fields commonly contain depressional areas that collect water from surrounding fields. These areas are thus typically higher in organic matter and should be sampled separately. Other unusual areas may be encountered, and these should be sampled separately, or at least the unusual characteristics should be noted in the project notebook.

Sampling must occur on both sides of rivers, as well as the bottom. This is especially true if contamination of the surface or subsurface is of concern. Even though it will not look like this can happen, over time all streams and rivers gradually work their way back and forth across the river valley and can thus both move and bury contaminants [16].

5.12. OTHER SAMPLING STRATEGIES FOR DIFFERENT FIELD SITUATIONS

5.12.1. The Grab Sample

In taking a grab sample, the question is what is being grabbed. Too often the grab sample is obtained in the excitement of first visiting a contaminated site. It is often taken without regard to its position in the contaminated area, the depth of sampling, the sampling tool, the labeling, or the type of container, and with no specific idea about the conditions of transport or analysis. For these and for safety reasons, grab samples are not advisable.

In addition, the grab sample is particularly inappropriate where contamination may occur in an episodic fashion. The most common example is the episodic contamination of water by agricultural chemicals, which occurs most frequently just before and after planting. If a water sample is taken before or after the planting no unusual amount of contamination may be found. It was not present yet or may later have been diluted to a natural or undetectable level. Either sampling at regular intervals or some method of taking a continuous sample, such as time-weighted average sampling based on solid-phase extraction, thus may be appropriate.

A specific example would be the contamination of groundwater by nitrate from agricultural fertilizer. The time to look for this is when fertilizer application is high and plant use is low. Grabbing a sample at some other time will not illuminate the problem [17,18].

5.12.2. Agricultural Sampling

Farmland represents the largest area sampled and analyzed per year. Typically, only surface samples 0 to 15 or 0 to 30 cm deep are obtained, and few if any deep samples are taken. The typical soil analysis includes pH, cation exchange capacity, organic matter content, and the amounts of calcium, magnesium, phosphorus, potassium, micronutrients, and sometimes nitrogen. Such analyses are inexpensive and rapidly done.

There are three types of agricultural fields, each requiring different sampling strategies (e.g., those planted to row crops, to grass, and to fields used for pasture). The most commonly sampled are fields with row crops (i.e., any field containing crops planted in discernible rows). Figure 5.10 shows a field planted to a row crop (corn) being sampled. In this type of field, fertilizers and other agricultural chemicals may be broadcast, or sprayed in such a way that the materials is spread evenly over

FIGURE 5.10 Sampling a field containing a row crop (corn).

the surface. If uniform application has been made, then random sampling is appropriate.

In the case of row crops, however, it is common for some fertilizers and chemicals to be placed in bands close to or in the row. When sampling this type of field, it is important to know how the materials were applied so that samples are not taken from areas likely to have high local concentrations of this material. This is particularly true if the analysis is for the applied material or one of its components. It is also important because the applied material may interfere or affect the analysis for other analytes.

A second type of field is one in which the fertilizers and chemicals are only added to the surface in a uniform manner. This would include both hay fields and grassed areas, such as golf courses, that are not pastured. A typical hay field is shown in Figure 5.11. Here random sampling is appropriate.

In the case of fields used for pasture, sampling is a little more complex. Here completely random sampling is not appropriate. Any place in which

FIGURE 5.11 An example of a hay field to which additions would be made uniformly.

animals congregate, such as around watering tanks or gates, will usually have high amounts of manure, and thus is not representative of the field as a whole. Also, when a sampling site is chosen it should not be covered by or be very near a pile of manure, as this will again constitute an area with unusual characteristics. Even if the manure is scraped off the top of the sample, organic matter and chemicals will have leached out of the manure and into the soil. The amount leached will depend on how long the manure has been in place and how much rain has fallen in the meantime. Both of these are unknowns and cannot be corrected for. A pasture field with animals grazing is shown in Figure 5.12.

The largest amounts of chemicals added to the environment each year are agricultural chemicals. In this group, lime ($CaCO_3$), which is usually spread evenly on the surface of fields, is used in the highest quantity in areas with acid soils. In all areas, nitrogen fertilizer is used in the greatest amount (second to lime on acid soils). Nitrogen is both uniformly spread on the surface of soil and as a band near the seed in row crops. There are also large amounts of pesticides—primarily insecticides and herbicides—used, while lesser amounts of fungicides and other agents to control nematodes, mites,

FIGURE 5.12 A pasture field with cows.

and so on are applied. In some cases, organic additions are applied, often in the form of animal manures, and these add not only plant nutrients, but also organic matter, microorganisms, and antibiotics given to farm animals. Table 5.2 gives estimates of the amounts of some amendments commonly added to agricultural fields.

5.12.3. Geological Sampling

Rock frequently has horizons that are horizontal, much like those in soil. Because of upthrusting of rock layers and intrusion of lava or a different rock layer into an already existing rock, however, there may also be layers that are not horizontal, and their slopes vary from $0°$ to $90°$. An intrusion of one rock into another is shown in Figure 5.13. Thus, it is possible to need to sample a layer that is vertical. In this case, a vertical sample may extend for considerable depth and the same material is sampled through the length of the sampler. It may be equally important in this case to sample adjacent rock vertically as well; however, the horizons in the adjacent rock may be horizontal and thus require an entirely different sampling procedure. A rock having horizontal layers is shown in Figure 5.14.

With steeply sloping layers, it is necessary to change sampling depth across the sampled area. It will be essential to have some identifying characteristic that will show the persons sampling that they are in the correct layer. A detailed record of the sampling depth across such an area is essential.

5.12.4. Environmental Sampling

Environmental sampling encompasses a wide range of activities that include not only air, water, and soil sampling, but also such features as noise

TABLE 5.2 Typical Agricultural Additions to Soil in Order of Decreasing Amounts

Fertilizer/pesticide	Chemical composition	Amounts added/ha[a]
Nitrogen	NH_4^+, NO_3^-	100–300 kg
Phosphorus	$H_2PO_4^-$	10–100 kg
Potassium	K^+	10–100 kg
Herbicides	Various	80–900 g
Insecticides	Various	200–2000 g

[a]Amounts do not represent any specific value but are intended to give a general idea of the amounts used. Specific information about amounts of herbicides and insecticides used can be found in *Crop Protection Reference*, New York: Chemical and Pharmaceutical Press, 2003.

FIGURE 5.13 A rock intrusion is evident in the middle of the picture.

pollution. Although different samplers are used (see below), the three-dimensional grid sampling concepts are applicable to all sampling, as are using GPS and GIS. In a detailed analysis of air, water, and soil, inorganic and organic gases, liquids, and solids, will be determined. These analyses frequently require a sophisticated laboratory and instrumentation that is expensive and time-consuming to operate, even when the contaminant is known [19].

Often environmental sampling benefits from time-weighted average sampling. Typically, placing an absorbent or cation or anion exchange polymer in contact with the media, soil, water, or air to be sampled accomplishes this type of sampling. The polymer is left in contact with the media for some time. It is then removed and the adsorbed material extracted from it and analyzed. The absorbent can be extracted with a solvent, or it may be heated to drive off the absorbed material. In the case of cation or anion exchange resins, the extracting solution must contain a cation or anion that will replace all cations or anions on the exchange resin.

The assumption is that the material adsorbed on the media is irreversibly attracted to the absorbent under the conditions in the field. It is also assumed that the absorbent does not become saturated during the

FIGURE 5.14   Rock showing various layers or horizons.

sampling time. If both these conditions are met, the amount of material on the sorbent divided by the time it is in contact with the media will give the average concentration of the contaminant in the media.

A word of caution is that there is another condition affecting the effectiveness of this method; there must be continuous, intimate contact between the media and the sorbent during the entire sampling period. If the sorbent is placed in a river and the river level alternately is above and below the sorbent, a representative sample for the time of sampling will not be obtained. This same question would arise if the sorbent is alternately in one layer and then in another layer of the media. Unless the time spent in each layer is known, the average amount of contaminant in the layers would not be accurately known.

A similar argument can be made where the sorbent is in a medium that is moving. An air-sampling sorbent that is alternately upwind and downwind from the contamination source will produce an analytical result that is difficult or impossible to interpret. The same thing would occur if the sorbent is placed in water having different currents or if the water is moving over it in different directions at different times.

In soil there is always some question about the contact between the adsorbent and soil. Coarse-textured and sandy soils have less contact than silty soils, while clay soils would have the most contact between the soil and the adsorbent. The question of contact thus must be adequately answered for both the absorbent being used or proposed for use and the soil it will be used in.

This type of sampling can be very effective where there is an episodic nature to the type of contamination being studied (i.e., contamination occurs at one time or over a limited time during the year). It is also important in cases in which the time of the occurrence of the contamination is not known. Placing absorbents in the media at various times and removing and analyzing them after a certain time interval allows for determination of the time of contamination [17,18].

5.12.5. Underwater Samples

Often sampling stops at or above the water table. Sometimes it is necessary to sample the soil or sediment at the bottom of a body of water, however. In this case, two additional complications often occur. First, the material is loose and tends to disperse while the sample is being taken. Second, the physical and chemical environment is very different at the bottom of a body of water from the environment into which it is to be placed, particularly in the analytical laboratory. These samples come from a high-pressure, anaerobic, reducing environment and are put into a relatively low-pressure, aerobic, and oxidizing environment. Because of these differences significant changes in components of interest can occur if precautions are not taken in obtaining and handling the sample (e.g., a sample containing large amounts of iron in an anaerobic, reducing environment is black; exposed to air, the sample becomes yellow, then red, and the iron precipitates).

A special sampler is needed when sampling the bottom of streams, rivers, ponds, and lakes. The sampler must retain the soil and its associated water and prevent contamination by overlying water. As with other sampling, a grid pattern can be used along with random sampling. The type of sampler needed can be constructed or obtained from commercial sources (see Appendix B) [20].

5.12.6. Other Sampling

To a large extent, sampling water and air is similar to what has been described for soil. Transect sampling should be carried out, including areas known not to be contaminated. This sampling must be three-dimensional (i.e., surface and subsurface water samples are taken). Atmospheric samples should be taken at various altitudes. The sampling should be as random as possible, but with attention to the different zones occurring in the area.

Differing zones are produced by strong currents of a river or in currents induced in a lake by a river, and they do not give a true picture of the state of the whole river or lake. Similarly, atmospheric samples taken at one altitude downwind from a pollution source do not give an adequate picture of the pollution. The same can be said for samples taken only upwind or away from entering water sources.

Here again sample size and uniformity of size are important, particularly if comparisons are to be made between samples. The same can be said for samples in which some regulatory criteria in terms of pollutant concentration are to be met [21–23].

5.12.7. Monitoring Wells

Where is the water in the field going? Not only is it necessary to sample the soil or the regolith, it is also essential to know where either rainwater or surface water goes when it enters from adjoining areas and exits a contaminated site. It is assumed that incoming water cannot be contaminated, but this must be verified. Incoming water is compared to both the water in the contaminated site and the water exiting the contaminated site in order to be sure of what is happening to the contaminant. Even if it seems that all the water in the area stays in place or just flows over the top of the soil it is still essential to keep track of it and its condition in and around the contaminated site.

To keep track of water and its condition, contamination, and movement, it is essential that wells be installed at the contaminated site. The most logical location is downstream from the contamination. It is essential to also have wells on all sides and upstream, however. These wells need to be sampled frequently throughout the sampling and remediation in the area. Keep in mind that well monitoring is recommended and sometimes required, even if the material at the site is to be removed for remediation or cleaning at some other site. Also, such monitoring wells are monitored for a period of time after the site has been certified as being clean [24].

5.12.8. The Number of Samples Needed

Methods for estimating the total number of samples needed to obtain a representative sample are presented in Chapter 6.

5.13. SAMPLE HANDLING

Sample locations can be chosen before actually going to the field. When in the field it is sometimes tempting to move the sampling site if it is in the middle of a tuft of grass, on top of a rock, or under a layer of plant residues. As much as possible, the site should not be changed. Take the sample near to the plant or under the rock, or take plant residue, if at all possible. Rocks or plant material occurring in the sample when it is removed from the soil should be left with the sample rather than being picked out. To maintain sample integrity and for the safety of the sampler it is best not to touch the sample with bare hands. In all cases, samples should be moved *directly* from the sampler to the sample container. They should not be handled and should not be removed, placed someplace, and then moved again into the container [12,13].

Cleanliness is also important. During sampling gloves will often be worn, and these must be changed between sampling events, particularly if the samples are heavily contaminated. Also, the sampling instrument needs to be cleaned between sampling events. This may involve washing or simply wiping the sampler.

5.14. SAMPLING USING GPS

Both the transect and detailed sampling sites are recorded using maps produced during the presampling phase. These can be recorded using either UTM or latitude and longitude. Using the UTM system or longitude and latitude, the notation for the position of the sites will take the form given in Table 5.3.

It is highly recommended that sampling sites always be given one of these two locator designations. During each sampling exercise a reference table giving the sample numbers and the corresponding locator positions will be produced. Note that the designations are given as examples and are not meant as actual positions on the Earth's surface. The project notebook designations are similar to those given in the notebook example in Chapter 3. However, they are not intended to represent the same samples.

If digital orthophoto quarter quadrangle (DOQQ) maps are being used, the position of the samples can be either manually entered into a computer or downloaded from the GPS unit. Then, using ArcView or similar GIS software, a map showing the location of the sampling sites can be produced by the computer. Such a map is shown in Figure 5.3, and can be produced using a GPS unit and a DOQQ map. Note that the map has a grid on it and sampling is done on the basis of this grid. It might also be noted that the samples are taken from random spots in the grids.

TABLE 5.3 Examples of Sample Sites Locations Using Universal Transverse Mercator (UTM) and Latitude and Longitude Designations

Field notebook designation	UTM designations		Latitude longitude designations	
	Easting	Northing	Latitude	Longitude
AC12a122602	485^{400m}·E.	37_{04}^{300m}·N.		
AC12b122602	485^{500m}·E.	37_{08}^{310m}·N.		
AC23a010603			N 47^0 19.56'	E 101^0 42.84'
AC23b010603			N 47^0 19.57'	E 101^0 42.83'

Figure 5.3 is intended as an example of how sampling might be done, but it does not represent an actual sampling situation. Examples of different ways of designating the samples are also shown. In the lower right-hand corner, the sampling site is designated by two sets of three numbers. These are the last three digits of the longitude and latitude positions. The whole position is N39 26.456 and W83 48.157. Only the three last digits need be shown, however, because the first four digits do not change throughout the field. This designation lacks the initials of the person doing the sampling and does not give the location in the project notebook in which a description of the sampling site is located.

The next sampling site is labeled by the sampler's initials and the page in the project notebook on which the samples are described. This is sufficient if the project notebook page also gives the position in the field using longitude and latitude. The initials of the person doing the sampling and six numbers representing the latitude and longitude location of the sampling site identify the next sample. In the last case, the operator's initials, the page number in the project notebook (13b), and the longitudes and latitude position are given.

In this last case the full project notebook designation plus the longitude and latitude may be too big to fit on the map, and an abbreviated form may be called for. For example, sample designation AC13b579190 might be simplified to AC13b579 or simply 13b. This may seem awkward, but is both invaluable as work continues and time-saving.

In all of the above situations, a second set of samples taken from a similar uncontaminated soil in a similar location is needed as a comparison and control. It is important in taking this sample that it comes from a field that has not been (and is not now) contaminated by any process, especially in relationship to contamination occurring at the field of concern. Again, both surface and subsurface samples are needed for comparison.

5.15. DETERMINING SAMPLES SITES

Random places to be sampled can be obtained by throwing darts at a map of a field and seeing where they land. In the case of field sampling, we could, for instance, take the field shown in Figure 5.2, enlarge it, and throw darts at it. Each place a dart lands would be a place at which a sample would be taken. Note that if this were done no distinction would be made between different soil types. Another way would be to assign a number to all possible sample locations in a field and randomly pick and sample these locations. Because there are an infinite number of possible locations, this is not a practical approach. Neither of these procedures is typically used in field sampling.

Another approach would be to pick a spot at the edge of a field as a starting point for sampling. For example, point A in Figure 5.1 is chosen as the starting point. Two randomly picked numbers are used to represent paces west and north. (Random number generation will be covered in Chapter 6.) The respective paces are taken and a sample obtained. This is then repeated to obtain a second sample. If only positive random numbers are possible, then sampling would proceed north and west across the field, leaving many portions of the field unsampled. If, on the other hand, both negative and positive numbers are used, a more random sampling covering more of the field would occur.

Although the above cases are statistically valid, they are not generally useful or practical in a typical field-sampling situation. First, it is not possible to analyze a very large number of samples for each and every area of interest. Second, the size of the area to be treated may be controlled by other than pure sampling or statistical considerations. For instance, when soil plant nutrient levels are determined as a basis for the commercial application of agriculture chemicals, the size of the area sampled is determined by equipment size, capacity, and economics.

For contaminated fields, we are generally interested in differentiating between contaminated and uncontaminated areas. It is important to know when the level of contaminant is at or below some specified target level. When three samples from an area have analytical results showing that the contaminant is below the target level at the 95% confidence level, then the desired information has been obtained and further sampling may not be needed. (Confidence levels will be discussed in Chapter 6.)

For these reasons, a more systematic sampling scheme is often used. An imaginary grid can be set up to cover the field. At this point each grid can be randomly sampled, or a selected number of grids can be sampled on a random basis. Even fewer random samples can be obtained by specifying positions within each grid area in which samples are to be taken. In this case, these positions can be repeatedly sampled to determine the changes in contaminant or component concentration over time.

In the case of agriculture, grids large enough to be treated individually are usually sampled. Here the desire is to have an estimate of the amount of fertilizer to be added to reach a certain yield goal, but not to cause environmental harm. Also, if an adjacent grid requires more or less fertilizer that adjustment can be made as needed. In this type of sampling, the average amount of available plant nutrient present in each grid is important for making fertilizer application. It should be kept in mind that very precise and accurate applications rate cannot be obtained, and because of this it is not necessary to be able to distinguish very small differences.

When contaminated areas are sampled the desired sampling outcomes are different. Where in situ bioremediation is to be carried out repeated sampling of the area will be necessary. Here we wish to make sure that the contamination is not spreading and to know when the contaminant has fallen below a specified level in a specific area. In other cases, the soil in the field sample is to be removed and remediated in another location. In this case, sampling will only be done once, and so consideration of repeated sampling is not in question [15].

5.16. QUALITY CONTROL

There is always a great deal of concern regarding quality control and quality assurance. Does the analysis and reporting of the results accurately and faithfully represent the situation in the field or in the environment? Often samples are spiked; a precisely known quantity of contaminant is added to a control sample. Spiking should be done in the field after the sample is in its container. If this precisely known quantity has changed during transportation, storage, or handling, the results of analysis of samples are suspect.

A problem arises when the contaminant of concern is either highly toxic or volatile. It is undesirable to increase the amount of toxic material in a sample, both from the standpoint of the safety of the personnel adding the contaminant and the person doing the analysis. In such cases, a surrogate may be added to the sample. This is a compound that is not toxic (or is very much less toxic), but has the same or similar physical, chemical, and biological properties as the component of interest. The surrogate allows one to determine if there is any change in the sample between sampling and analyses without undue danger [25].

Liquid or solid contaminants or surrogate compounds added to soil or regolith materials do not cover them uniformly. Only analysis of the whole sample will give an accurate picture of the fate of the contaminant or surrogate. A subsample may give an erroneous result because of nonuniform distribution, which can occur even if the sample is ground, sieved, and mixed exhaustively.

5.17. LABELING

Labeling seems like a simple part of any sampling activity. Incomplete, unreadable, and missing labels cause error, however, as well as a great deal of wasted time and money.

5.17.1. Labeling a Sample

Labels must include reference to the field position from which the sample was taken and to the page in the project notebook on which sampling is described. Numbering of the designated sampling areas, as indicated in Figure 5.9, is done sequentially. The sample areas below the surface are indicated by Greek letters. In this instance, one would know that sample 15β is the second sampling area below surface area 15. This gives an unequivocal designation and location to this sample. The question might be asked as to why Greek letters are used when the Roman alphabet will do. What most researchers find is that they have use for the alphabet in designating other aspects of the sample. We therefore save our alphabet for these uses and use the Greek alphabet for designating subsurface sampling areas. A sample number must be unique, but allow for easy lookup when needed.

The project notebook page on which the sampling is described is on the label, which must also show the date and time sampling was performed as well as who took the sample. Additionally, where it is to be stored and shipped must be shown, as well as the shipping route.

There are other considerations in deciding on how to label a sample. First, during handling, will the samples be subject to wetness, even if only from workers' hands? The answer is always yes. This means that any numbering or other designation that can be removed with water will be removed, or at least smeared until it is no longer recognizable. The first step is thus to use only "tested" permanent markers (e.g., a test sample number is applied to the sample container when it is dry, and then it is wiped with water, soap and water, etc.).

Another strategy is to put the sample label in several places on the sample container. Thus a copy of the label can be put on the bottom of the sample container, since people will not be rubbing against the bottom as much as the sides. The drawback is of course is that the bottom may be more moist that the top, and this may lead to smearing. The bottom may also rub by sliding during transport. If the sample container is one that is folded over and sealed by holding the folded area in place with ears, then the top that is folded over is a good spot for a copy of the label. Other good labeling places will exist on other types of sample containers.

This is not to indicate that the label should not be placed in some prominent location. It should. It must occur prominently on the side on the container and must contain all the information indicated above and information as to the organization requiring the sampling or paying for the analysis. Other pertinent data, such as phone numbers and e-mail addresses, should be included as necessary. This label is for convenience and efficiency in processing samples. Sample numbers in other places are for safety and need only be the sample number. Serious consideration should be given to including bar coding on the label. (See Chapter 3.) Bar codes can be placed on all samples and a duplicate placed in the project notebook and the chain of custody document. (See Chapter 8.) Not only does this simplify keeping track of samples, it will also decrease the chance of error in reading and reporting sample labels [26,27].

5.18. CONCLUSIONS

Sampling starts when all the preliminary data have been gathered and the equipment has been assembled. A map of the area to be sampled is prepared, giving its position on the Earth's surface. Once this is done, the next step, assuming ground-penetrating radar (GPR) and remote sensing are not to be used or have already been used, is to do a transect sampling, including contaminated and uncontaminated areas. With these data a detailed sampling plan involving gridding and labeling the areas to be sampled is completed. The sampling plan may be random or may include a combination of random and nonrandom components. As the samples are taken they are placed in suitable containers for transport and analysis and given appropriate labels relating them to the field notebook and position on the Earth's surface. Additional tools such as monitoring wells and GIS may also be used to obtain a more complete picture of the sampled field.

QUESTIONS

1. Describe the differences between a grab sample, transect sampling, and detailed sampling.
2. What are two coordinate systems used in GPS?
3. Describe all the information that a label should contain and where it should be on the sample container. In answering this question do not limit yourself to this chapter alone.
4. What is the accuracy of GPS and remote sensing?
5. What is unusual about samples that are taken deep in water or soil that is saturated with water? What precautions need to be taken with these samples?
6. There are three common types of samplers. Name the sampler types and describe the situations to which they are best suited.
7. Explain how and why Greek letters might be used in labeling a sample.
8. The spectral range for satellite imagery is over what range of the electromagnetic spectrum?

9. Explain which sample container would be used for a soil sample containing a heavy metal contaminant and one containing a volatile organic carbon (VOC) contaminant. Explain the bases of your choices.
10. Explain grid sampling.

REFERENCES

1. Wagner G, Desaules A, Muntau H, Theocharopoulos S, Quevauviller P. Harmonization and quality assurance in pre-analytical steps of soil contamination studies—Conclusions and recommendations of the CEEM soil project. Sci Total Environ. 2001; 264:103–117.
2. Créptin J, Johnson RL. Soil sampling for environmental assessment. In: Carter MR, ed. Soil Sampling and Methods of Analysis. Ann Arbor; MI: Lewis, 1993:5–18.
3. Patterson GT, Site description. In: Carter MR, ed. Soil Sampling and Methods of Analysis. Ann Arbor; MI: Lewis, 1993:1–3.
4. Petersen RG, Calvin LD. Sampling. In: Klute A, ed. Methods of Soil Analysis. Part 1—Physical and Mineralogical Methods. 2nd ed. Madison, WI: American Society of Agronomy and Soil Science Society of America, 1994:33–51.
5. Letham L. GPS Made Easy. Seattle: Mountaineers, 2001.
6. Conyers, LB, Goodman D. Ground-Penetrating Radar: An Introduction for Archaeologists. New York: Rowman & Littlefield, 1997.
7. Lunetta RS, Elvidge CD. Remote Sensing Change Detection: Environmental Monitoring Methods and Applications. Chelsea, MI: Ann Arbor Press, 1998.
8. GSFC Earth Sciences (GES) Distributed Active Archive Center. http://xtreme.gsfc.nasa.gov/.
9. Earth Observing System. http://edcimswww.cr.usgs.gov/pub/imswelcome/index.html.
10. Korte GB. The GIS Book. 5th ed. Albany, NY: OnWord Press, 2001.
11. Description and Sampling of Contaminated Soils: A Field Pocket Guide. EPA 625–12–91–002. Cincinnati, OH: National Center for Environmental Publications, 1991.
12. Bates TE. Soil handling and preparation. In: Carter MR, ed. Soil Sampling and Methods of Analysis. Ann Arbor; MI: Lewis, 1993:19–24.
13. 5th ed. Sample Handling and Transmittal Guide. http://www.rdc.uscg.gov/msl/downloads1.html.
14. Dent D, Young A. Soil Survey and Land Evaluation. Boston: Allen & Unwin, 1981.
15. Koryta J, Stulik K. Ion Selective Electrodes. 2nd ed. Cambridge: Cambridge University Press, 1984.
16. Hollands KR. Comparing Topography Soil Sampling with Other Known Precision Ag Methods. http://www.sbreb.org/96/soilmgmt/96p98.htm.
17. Battaglin WA, Hay LE. Effects of sampling strategies on estimates of annual mean herbicide concentrations in midwestern rivers. Environ Sci Tech. 1996; 30:889–896.

18. Koziel JA, Noah J, Pawliszyn J. Field sampling and determination of formaldehyde in indoor air with solid-phase microextraction and on-fiber derivatization. Environ Sci Tech 2001; 35:1481–1486.

19. Brady NC, Weil R. The Nature and Properties of Soils. 12th ed. Upper Saddle River, NJ: Prentice-Hall, 1999:656–658.

20. Fisher MM, Brenner M, Reddy KR. A simple, inexpensive piston corer for collecting undisturbed sediment/water interface profiles. J Paleolim 1992; 7:157–161.

21. Hess-Kosa K, Hess K. Indoor Air Quality: Sampling Methodologies. New York: Lewis, 2001.

22. Compilation of EPA's Sampling and Analysis Methods. 2nd ed. Keith LH, ed. New York: Lewis, 1996.

23. National Field Manual of the Collection of Water-Quality Data. Techniques of Water-Resource Investigations. Book 9. Handbooks for Water-Resource Investigations. http://water.usgs.gov/owq/fieldmanual/ (search for Book 9).

24. Monitor Well Design, Installation and Documentation at Hazardous and/or Toxic Waste Sites (Technical Engineering and Design Guides as adapted from the U.S. Army Corps of Engineers, no. 17). American Society of Civil Engineers. Reston, VA: ASCE Press, 1996.

25. Klesta EJ Jr, Bartz JK. Quality assurance and quality control. In: Klute A, ed. Methods of Soil Analysis. Part 3—Chemical Methods. Madison, WI: American Society of Agronomy and Soil Science Society of America, 1996:19–48.

26. Holkham T, Holkman T. Label Writing and Planning. New York: Aspen, 1996.

27. Bushnell R, Dooley T. Bar Code Compliance Labeling for the Supply Chain: How to Do It. Surf City, NJ: Quad II, 2000.

6

Statistics

Statistics provide very powerful tools for analyzing data, including tools for analyzing sampling activities. Some of these common tools are determining the number of samples needed, standard deviation, regression analysis, and extraneous values, and predicting the component values between sampled points. In all cases calculators or computers equipped with standard software, including spreadsheets, can be used to do statistical calculations. In order to understand what information the statistic is providing, however, and to be able to explain this to others, it is important to know how the statistic is calculated. This is also important when trying to understand what others are trying to explain using statistics.

It is common to hear people talk about median, mean, average, one standard deviation, and so on. When such conversations occur, are the speakers using the terms correctly? How does what they are saying affect your sampling plan? On the other hand, can the statistics being used in the sampling plan be explained to others?

There are a great many symbols used in statistics. It is important to be able to keep these in mind while studying and interpreting statistical results. Table 6.1 gives some of the most commonly used symbols and the terms they represent. These will be particularly important, whether calculation of statistics is done by hand or using a computer, because once the computer has done the calculation the experimenter must still interpret the statistical results.

Statistics can be calculated by hand; however, in most cases calculations will be carried out using computers and more or less complicated statistics programs. Although for simple calculations a handheld calculator can be used, in all cases a computer will be preferred. In the case of geostatistics the calculations are never done by hand or handheld calculators because the equations are too complex. Keep in mind that some statistics programs are difficult to learn and use, while others are not. In addition, some statistics programs leave the experimenter in charge; some take charge.

For example, some statistics programs require setting up the statistics or the sampling plan before obtaining or entering the data. Other programs allow the input of the data and subsequent application of various statistical calculations. Because it is not always clear what information will be needed or which statistic might need to be applied, the latter approach is usually preferable. In Figure 6.1, after the data are entered it is found that two data points are missing (X in A2 and Y in A6). If all the statistics had been set up beforehand [e.g., the number of samples and the degrees of freedom (n and $n-1$)], the whole setup would have to be changed because of the missing data points.

If the data are entered and the statistics subsequently calculated, two possible adjustments could be made. First the data could be calculated

TABLE 6.1 Important Statistical Symbols and Their Meaning

Symbol	What it is	How is it obtained
μ	Population mean	Add all data points and divide by the number of data points in the population.
$\bar{x}$	Sample mean	Add all sample data points and divide by the number of sample data points taken.
s^2	Sample variance	Obtained by taking the difference between the sample mean and each sample and squaring and dividing by $n-1$.
s	Sample standard deviation	Take the square root of s^2, the sample standard variance.
H	Hypothesis	The experimenter determines the null and alternate hypothesis.
t	t statistic	Obtained from table of t-values.
F	F statistic	Obtained from table of F-values.

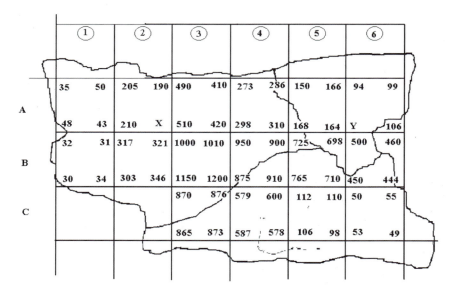

FIGURE 6.1   Total petroleum hydrocarbons (TPH) levels in various areas of a field.

without the missing data points using a smaller n and $n-1$. On the other hand, it might be decided to calculate the missing data points using kriging (described below). In either case, being able to manipulate the data after they are entered would facilitate these changes.

6.1.   RANDOM NUMBERS

In Chapter 5 random numbers were used to determine where samples were to be taken. How does one obtain random numbers? A simple, unsophisticated way to do this is to place little pieces of paper with numbers on them in a hat and pull them out one by one. Random numbers can be more easily generated with many handheld calculators and with all computers. Normally there will be a function called rand that will return random numbers. Sometimes it will be necessary to multiply or truncate the numbers obtained to produce numbers in the range you wish to have.

As long as the numbers are used as given by the program they are random, even if they are multiplied by 10 or truncated to only one decimal place or a whole number. Table 6.2 is an example of random numbers generated by Excel using "= rand()*10." The result of this operation is the random numbers given in column 1. The truncated and rounded-up

TABLE 6.2 Random Numbers Generated by
Excel Using = rand()*10

Rand()*10	Rounded up
8.879528391	9
1.622723374	2
5.014714332	5
3.285093315	3
1.682229083	2
3.64113318	4
6.351850918	6
6.660649484	7
6.174987005	6
4.024443714	4
4.247216289	4
9.915704505	10
6.332125286	6
8.592222671	9
2.90158629	3
3.030321588	3
0.448112927	0
9.875606847	10

Note: Column 2 is the random numbers rounded to
whole numbers.

numbers corresponding to these random numbers are given in column 2.
Any time random numbers are needed this is the method best suited to
obtaining them.

6.2. VARIATION

When dealing with environmental samples it is common to have variation
that is much larger than the chemist or mathematician would like. For some
samples a variation of $\pm 1\%$ might be acceptable; however, with
environmental samples a variation of $\pm 10\%$ may represent a very accurate
measurement.

For example, a soil core 1.9 cm in diameter was taken and the pH of
each of three subsamples was measured, first with the fresh sample, then the
sample after it was dried, and again after the sample was dried and mixed. In
each case subsamples were mixed with water (1:1 ratio), and the pH
determined using a standardized pH meter. Table 6.3 gives the soil pHs

TABLE 6.3 The pH Measurements of Three Replicate Samples from a Soil Core in the Original, Air-Dry, and Air-Dry Mixed Conditions

Sample	Fresh wet sample	Dry-sieved sample	Dry-sieved mixed
AC12a	5.30	5.32	5.23
AC12b	5.03	5.41	5.28
AC12c	5.01	5.27	5.30
Mean or average	5.11	5.33	5.27
Standard deviation	0.16	0.07	0.04

Note: Unpublished data from senior research project of Luke Baker, Chemistry Department, Wilmington College, Wilmington, Ohio.

found. The pH values in the second column were obtained after the core, which was sealed in a plastic bag, was crushed by hand and mixed while still in the bag. The third column is the same samples after air-drying and sieving (but no further mixing). The fourth column is the same samples after mixing on a mixer overnight.

It will be noted that variation in the first set of measurements is quite large. Even though the first measurement appears to be in error, including it gives a better estimation of the mean than excluding it. Drying the sample leads to a decrease in acidity, while sieving and mixing had little effect on the pH but decreased the variability. The difference between a pH of 5.11 and a pH of 5.33 is so small that it would cause no change in a remediation plan. Later in this chapter these pHs will be compared and their relationship to each other examined [1].

6.3. POPULATION

In sampling groups of people or fields of corn the concept of sampling a population is self-evident. Air, water, and soil appear to be continuous media, however, and not related to populations at all. In this case each sample is considered a member of a population. One way of looking at this is to ask if the samples are part of a group of samples that are the same or if they are from a different group of samples.

Soil scientists do define a soil individual; that is, a volume of soil that has well- and precisely defined characteristics. Such a soil, which is 1 to $10\,m^2$ and 1.5 to 2 meters deep, is called a pedon, and a group of pedons in a field would be called a polypedon. It is rare that a field to be sampled would consist of only one type of soil or pedon, however. When soils are mapped, as in a soil survey (see Figure 1.2 in Chapter 1), the different soils noted are not technically pedons, but soils of similar characteristics called mapping units. Soil sampling thus is not sampling individuals in the soil science sense.

Both the atmosphere and hydrosphere are also seen to be continuous and homogenous. When they are sampled, however, each sample is considered to be a member of a population of samples, just as soil samples are.

Often three soil cores will be crushed and mixed to provide a sample for analysis. This sample is also considered to be a member of a population of soil samples. This is true as long as it is analyzed as one sample. It is expected that these samples will show the same characteristics as any other sampling of this population or area; that is, plotting the frequency of a characteristic versus, the number of individuals showing that the characteristic will produce a standard, symmetrical bell-shaped curve. (See Figure 6.2.) A characteristic that produces such a curve when measured and graphed is a normal random variable. There are many similar curves with different absolute shapes that are considered bell-shaped curves. The shape of the curve and the fact that the mean and median are the same shows that its probability distribution is a normal distribution.

This same standard distribution argument can be made for other samples. A series of air, water, or regolith material samples are all expected

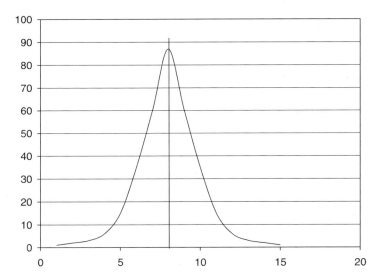

FIGURE 6.2 A bell-shaped curve. The line shows the median and the mean for this set of data. (A true bell curve is calculated using the formula $1/e^{x^2}$.)

to have this standard normal distribution characteristic. In all cases this assumes that a large population is sampled; in most field sampling situations, however, neither a large sample population nor a large number of samples is feasible for reasons of either time or economics. Unless we know otherwise, the assumption is that the population is a normal distribution and that the sample taken is representative of that normal distribution.

The shape of the curve will be significantly different for a population if the mean and median are different. This will produce a bell curve that is not symmetrical (or it is skewed). This would mean that on either side of the mean the curve would be different. Such a situation is illustrated in Figure 6.3.

6.4. HYPOTHESIS

One of the first steps in using statistics is to develop a hypothesis that is used in interpreting the results of the statistical analysis. Typically the hypothesis is that two populations or sets of data represent two different populations or are from the same population. In field sampling we are often interested in using statistics to determine if an analyte is above or below some cutoff level. Another way in which hypotheses are used is to determine if two areas being sampled have the same or different levels of contamination. Two

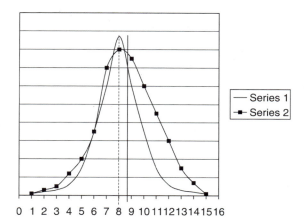

FIGURE 6.3 Graph showing a rightward-skewed distribution (series 2). The dashed line is the median and the solid line the mean.

adjacent areas having the same level may be sampled as one unit rather than two, saving time, effort, and money.

Examples of hypotheses are given in Table 6.4. The first two hypotheses state that the population sample is equal to (in the first case) or not equal to (in the second case) the sampled population, respectively. The second two equations state that the population is larger than the sample population and smaller than the sample population, respectively. Hypotheses are extremely valuable and powerful tools in any research. Whether or not a hypothesis is correct is immaterial, because no matter which is the case, valuable information about the samples and the populations being studied is obtained. Hypotheses should never be looked at as being correct or incorrect.

Looking at the pHs in Table 6.3 we might have a hypothesis that the pHs are all from the same population (i.e., all the samples are the same). This represents the first hypothesis in Table 6.4 and is the one we will use to test whether or not the pHs are the same. Another way to state this is that the method of determining the pH of soil does not change the pH obtained. This hypothesis will be tested later in this chapter [2].

6.5. MEDIAN AND MEAN

Two essential pieces of information about a set of data is to determine the central tendency, which is the median or the mean, and the standard deviation. These numbers form the basis for all other calculations. The median is found by arranging the data in ascending or descending order. If the number of data points n is odd, the middle number is the median. If the number is even, the median is the average of the middle two numbers. There are other ways of calculating the central tendency, such as the average (same

TABLE 6.4 Hypothesis Used in Statistical Analysis Where μ_0 is the Specified Population Mean and μ Is the Actual Population Mean

Hypothesis	Meaning	t statistic
$H_0 : \mu = \mu_0$	Populations are the same.	Two-tailed t
$H_0 : \mu \neq \mu_0$	Populations are not the same.	Two-tailed t
$H_0 : \mu_0 \rangle \mu$	μ_0 is larger than μ. Sampled population is smaller than specified population.	One-tailed t
$H_0 : \mu_0 \langle \mu$	μ_0 is larger than μ. Sampled population is larger than specified population.	One-tailed t

as mean) and mode. The value of the median, however, is that 50% of the values fall above it and 50% below.

The other central tendency calculation is the mean, which is the same as the average and is used to calculate other statistics. To obtain the mean of a group of numbers, the numbers n are added together to find the total. This total is divided by the number of numbers and the result is the mean. If the data set is large and evenly distributed, then all the measures of central tendency will produce the same result. When the median and mean are not the same, the data are said to be skewed. A graphical representation of such data is given in Figure 6.3. If the mean is to the right of the median, the data are said to be rightward skewed, as shown in Figure 6.3. If it is to the left, it is said to be leftward skewed.

When discussing statistics it is important that everyone knows which measure of central tendency is being used. As shown above, the relationship among the median, mean, and the other values of central tendency can provide useful information about the nature of the data.

As an example of the calculation of the mean, take the data from area A4 on the map in Figure 6.1. These are the analytical results for the concentration in ppm of total petroleum hydrocarbons (TPH) in the four surface core samples taken in this location. There are four analytical results—273, 286, 298, and 310 ppm, respectively. The mean is represented by the term $\bar{x}$ which is called x bar. This term is found by adding numbers and dividing the sum by n (n is the number of numbers)

$$\bar{x} = \frac{x_1 + x_2 + x_3 + x_4}{n}$$

or taking the values for area A4 in Figure 6.1 would give

$$\bar{x} = 292 = \frac{273 + 286 + 298 + 310}{4}$$

This number is found by rounding up.

The median for these same data is also 292 ppm. This is obtained by adding the central two numbers (i.e., 286 and 298) and dividing by 2.

Carrying out the same calculation for B4, the mean (i.e., $\bar{x}$) is found to be 909 ppm (rounded up). Even though the means are different from those for A4 they do not provide enough information to tell if the two areas have different amounts of TPH or not. For this determination calculation of the standard deviation, shown below, is needed.

The average pHs given in the next to bottom row of Table 6.3 were also obtained in this way. In this case each of the pHs shown was obtained by finding the mean of the three determinations of the pH of the soil core shown in the column above it [3].

6.6. STANDARD DEVIATION

The first step in determining if the two average TPH levels or the average pHs calculated above are different or not is to first calculate the sample variance s^2 and the standard deviation s. The first step in this process is to determine how far away from the mean each of the data points is. To obtain this information each data point is subtracted from the mean. From above the mean $\bar{x} = 292$ and

$$273 - 292 = -19 \qquad 286 - 292 = -6 \qquad 298 - 292 = 6$$
$$310 - 292 = 18$$

Because this gives us a mixture of positive and negative numbers the mean of these numbers is either 0 or very close to 0, and thus gives little information about the variability.* To make the calculation more meaningful the differences are squared, which makes all the numbers positive. Thus

$$(-19)^2 = 361 \qquad (-6)^2 = 36 \qquad 6^2 = 36 \qquad 18^2 = 324$$

A more useful mean is then determined for these squared numbers:

$$\frac{361 + 36 + 36 + 324}{4} = 189$$

This calculates the population variance (δ^2). In mathematical terms what has been done in the numerator is shown by the term $\sum_{i=1}^{n} (x_i - \bar{x})$.† This term says that the difference between the mean $(\bar{x})$ and the value is

---

* If no rounding is used in determining the mean, the sum of the deviations from the mean will be 0.

† The letter Σ is used to represent a summation. It tells us that the group of numbers represented by $(x_i - \bar{x})^2$ are added together. The subscript i indicates that there are several different numbers. In this case the first number, 273, is substituted for x_i and the calculation is carried out. The second number is then similarly taken, then the third, until all the numbers have been substituted for x_i and the calculation carried out. Finally, all the numbers so obtained are added together. This is summarized in the following equation:

$$\sum_{i=1}^{4} (x_i - 292)^2 = (273 - 292)^2 + (286 - 292)^2 + (298 - 292)^2 + (310 - 292)^2 = 757.$$

calculated then squared and the numbers added together to give the answer (numerator). The final step is to divide this sum by n. This type of calculation is programmed into most handheld calculators and computers. It is such a common term that it is part of the standard keypad of many handheld calculators and the tool bar of Excel.

In calculating the sample variance $n - 1$ (or in this example, 3) would be used as the divisor. The number $n - 1$ is sometimes called the degrees of freedom. The reason for using this number rather than n will not be covered here, but can be found in all standard statistics books, such as those given at the end of this chapter. When the sum of the deviations is divided by $n - 1$ the number obtained is frequently represented by the symbol s^2, which is called the sample variance. The sample variance is used as an estimator of the total population variance. The mathematical formula for carrying out the sample variance calculation is

$$s^2 = \frac{\sum_{i=1}^{n} (x_i - \bar{x})^2}{n - 1}$$

A shortcut or direct way to carry out this calculation would be to use the following equation:

$$s^2 = \frac{\sum_{i=1}^{n} x_i^2 - (\sum_{i=1}^{n} x_i)^2 / n}{n - 1} = 252$$

Although this equation looks complicated and is, using a computer it can easily be programmed into the spreadsheet and used to make calculations. In most cases the spreadsheet will already have the capability to calculate the standard deviation, and so entering the equation is not necessary.

To calculate the standard deviation, represented by s, the square root of the variance (i.e., $s = \sqrt{s^2}$) is determined. Thus

$$s = \sqrt{s^2} = \sqrt{252} = 15.87$$

At this point we have the standard deviation, which is 15.87. This is also expressed as and has the units ppm, just as the original data do. This is interpreted to mean that the actual value is most likely to be somewhere between $292 + 15$ or 307 and $292 - 15$ or 277. The smaller the standard deviation, the better the number or the more precise the measurements.

The standard deviation can be used in many ways. In field sampling it is often used to determine if a portion of a field is above or below a predetermined level of some component. In the case of plant nutrients it may be used to tell if the nutrient is within a desired range.

Another way to look at this is to ask how many measurements are within one standard deviation of the mean. In this case one standard deviation is 15 ppm or ± 15 ppm; that is, how many or what percentage of measurements are within one standard deviation above and below the mean? Typically it would be expected that 68% of the measurements would fall within one standard deviation. In the example used here a very small number of measurements (only four) are used, and only 50% are within one standard deviations (i.e., between 307 and 277).

In Table 6.3 the standard deviation for the pHs obtained for each sample is given. The second and third samples are within one standard deviation of each other, while the first sample is outside one standard deviation of the other two samples.

6.7. DIFFERENT OR SAME POPULATION?

There are a lot of statistical tools appropriate to the comparison of two sample means $\bar{x}$. Most of these are appropriate for a more complicated analysis. When only a relatively simple analysis is needed, the statistic used is called the t statistic or student t. This is the statistic that will commonly be used in almost all field and environmental sampling.

Is the average TPH concentration in area A4 the same or different from the concentration in sample area A3? The first step is to decide on a null hypothesis. In this case let the null hypothesis for A3 be $H_0 : \mu = 457.5$. The alternate hypothesis is that $H_0 : \mu < 457.5$. For this calculation the degrees of freedom are 3 (i.e., $n - 1$ or $4 - 1$). Is the μ of A3 smaller or larger than the mean of A4, which is 291? To determine this, a t-value for these data is calculated using the two-sample t-test using the following equation:

$$t = \frac{\bar{x}_1 - \bar{x}_2}{\sqrt{\frac{s_1^2}{n_1} + \frac{s_2^2}{n_2}}} = \frac{457.5 - 291.75}{\sqrt{\frac{(49.92)^2}{4} + \frac{(15.88)^2}{4}}} = \frac{165.75}{26.19} = 6.33$$

The $\bar{x}$ for the area A4 is 291.75 and the μ_0 for A3 is taken to be 457.5.

How confident can one be that the two means are the same or different? Looking at Table 6.5, there are confidence intervals for values of t. If a confidence interval of 95% is chosen, this means that there is a 95% chance that the samples are the same if the t-value is less than that given in Table 6.5. Looking at Table 6.5 it is seen that t at the 95% confidence

TABLE 6.5 Abbreviated Critical Values of t and Q

$n-1$	Percentage confidence intervals for t				90% confidence interval for Q
	90%	95%	99%	99.5%	$Q_{0.90}$
1	3.078	6.314	31.821	63.657	—
2	1.886	2.920	6.965	9.925	—
3	1.638	2.353	4.541	5.841	0.941
4	1.533	2.132	3.747	4.604	0.765
5	1.476	2.015	3.365	4.032	0.642

interval is found to be 2.353. Because this value is less than the calculated t the null hypothesis is rejected. This means that the two means are different and represent two different populations.

This seems like an obvious outcome because the means are so different. The question could be asked if areas A1 and C6 are different or the same. Again a null hypothesis is needed. From Table 6.4 we let the null hypothesis be $H_0 : \mu = 45$. (The data are shown in Table 6.6.) Then the calculation of t becomes

$$t = \frac{51.75 - 44}{\sqrt{\frac{(6.683)^2}{4} + \frac{(2.753)^2}{4}}} = \frac{7.75}{3.61} = 2.14$$

TABLE 6.6 TPH Means and Standard Deviations

Sample number	Mean (rounded)	Standard deviation[a]
A1	44	6
A3	455	49
A4	292	15
A5	165	8
B1	31	1
B2	324	17
B3	1090	100
B4	905	31
B5	717	29
B6	455	25
C5	108	6
C6	51	2

[a] Truncated

This *t*-value is smaller than the *t*-value for the 95% confidence interval, and so the two are most likely (95% confident) to represent the same populations. If the confidence interval was chosen to be 99.5% (see Table 6.5) the hypothesis would also be true, and the two areas would be considered to be sampled from the same populations. However, the confidence interval should always be chosen before the statistics are calculated, should be used consistently, and should *not* be adjusted afterwards.

Keep in mind that the smallest number of measurements that can be treated statistically is two. Less that this and *n*-1 becomes 0. In realistic terms the smallest number most people like to work with is three. This is a compromise between a large number of samples (which would be prohibitively expensive, both in terms of time and money) and so few samples that no inferences can be made.

The *t* statistic can also be used to determine if the true TPH level in any sampling area is above or below the cutoff level. Looking at areas C3 and C4 it is obvious that the levels are above the cutoff level, assuming this to be 500 ppm TPH, and that the sampling plan needs to be amended to sample unsampled adjacent areas. You might wish to check areas A3 and A6 to be sure that they are below the cutoff level. To do this a different *t* statistic is used. This is the one-sample *t* statistic. This calculation would be carried out using the equation below, where $\mu_0 = 500$.

$$t = \frac{\overline{x} - \mu_0}{s/\sqrt{n}}$$

Also, areas A2 and A6 are missing data points, which need to be determined or estimated using kriging. (See below.)

In the examples given here sample means and hypotheses are evaluated individually. There are tools and methods for evaluating multiple sets of data points at the same time. Analysis of variance is one of these methods and will be discussed below [4]. For the data in Table 6.3 the *t*-test can be used to determine if the results from the various tests are the same or different.

6.8. EXTRANEOUS VALUES

In all field sampling the results of analysis will contain values that are or appear to be extraneous; that is, they appear to be larger or smaller than expected or are larger or smaller than the general trend in the data would indicate is expected. In some cases the term outliers may be used to mean the same thing. Another interpretation is that extraneous values are those that do not come from the population, while outliers do come from samples from the same population.

The first place to start in trying to determine if a value is extraneous is the project notebook. Is there any indication that there is anything unusual about this sample site or the sample when it was taken—any color variation, any indication of compaction, any unusual wetness, or any other observation that might indicate that the sample was different? A second place to look would be the chain of custody. Did anything unusual happen to the sample during transport or storage? Was there a change in temperature, was the box dropped, was the sample stored for an unusually long period of time? Did anything unusual happen at the laboratory before the actual analytical procedure was performed?

If some unusual situation occurred with the sample in question, the analytical data from that sample can justifiably be discarded. It should be noted, however, that if any other samples showed the same unusual characteristics or suffered the same conditions and the analytical results appear to be normal, then these conditions *cannot* be used as a reason to discard data.

If there is a mistake made during the analytical procedures, then it is reasonable to expect that the data would be thrown out and not reported to the end user. We would therefore not expect the actual analysis to be questionable.

When all other sources of variability are ruled out, the data can be checked using statistical methods. Because of the large standard deviation in sample A3, it might be suspected that one of the values is in error. The first step in estimating this is to find the range of observations. This is the difference between the largest and smallest value, and is represented by w, which is calculated as follows:

$$w = 510 - 410 = 100$$

Another way to think of this value is that it represents the dispersion of the values. A simple statistic for deciding if a value is extraneous is the Q test.

$$Q = \frac{|x_2 - x_1|}{w}$$

The distance between the questionable value and its nearest neighbor is determined and divided by w. If we were testing to see if 510 is an extraneous value in sample area A3, this would be $510 - 490$, which equals 20.

Calculating Q

$$Q = \frac{510 - 490}{100} = \frac{20}{100} = 0.2$$

Because this value is less than the $Q_{0.90}$ or the 90% confidence interval (see Table 6.5) the value is not considered to be extraneous and is kept in the calculations.

In Table 6.3 the first pH value would appear to be extraneous or an outlier. Looking at the mean pH values, the inclusion of this value is correct. After drying, sieving, and mixing it is seen that the average calculated with the first value included (5.11) is much closer to the one obtained after drying and mixing (5.27) than it would be if that value is excluded from the calculation (i.e., 5.02).* This points out the importance of not excluding values too hastily. We could also try to test this by finding w, which would be $5.30 - 5.01 = 0.29$. Then Q can be calculated.

$$Q = \frac{5.30 - 5.03}{0.29} = \frac{0.27}{0.29} = 0.93$$

From Table 6.5 we see that there is no value given for $3 - 1$ or $n = 2$; thus we cannot use this method to make this determination.

Extreme care should be taken in disregarding or discarding data. It is very easy to produce skewed or biased data by discarding data without a firm, unbiased basis for their elimination. The best reason for discarding a data point is having recorded data showing how, why, and where an error has occurred [5].

6.9. HOW MANY SAMPLES?

Another area of concern is to determine how many samples need to be taken. The concern is both from the standpoint of obtaining an accurate description of the situation occurring in the field and in minimizing the cost of sampling and analysis.

---

* There are other methods of estimating the validity of a measurement. They will not be covered here, but can be found in the references.

A simple method of calculating the needed number of samples is given by the equation

$$n = \frac{t^2 * s^2}{D^2}$$

In this equation $n =$ the number of samples to be taken. The t is from the t-table (Table 6.5) for the confidence interval needed. The variability in the samples is represented by s^2, which is the variance. D is the acceptable variability in the mean estimate. There are two ways of handling this equation. One is to assume that the area is to be remediated to no or 0 contamination. The other is to define the level to which the contaminant is to be reduced.

This relationship can be handled in several ways, depending on the sampling needs. One could specify a value for n and calculate either s^2 or D^2. On the other hand, s^2 and D^2 may be known or assumed and n calculated. One way to estimate s^2 is to use $(R/n)^2$. Here R is the expected range in the sampling, or in this example $(R/4)^2$ is an estimator of the true s^2 for area B5.

The applicability of this equation is complicated by several factors. One is that the variance or variability may be different at different places or depths in a field. If materials are to be removed, only the number of samples needed to determine the boundary between an acceptable and unacceptable level of a component needs be calculated. If this is done and the cutoff is 500 ppm, then in B5 we could use $724.5 - 500 = 224.5$ as R. This would then be used to calculate the number of samples to be taken. In this case, $(224.5/4)^2$ or $(56.12)^2$ would be used as an estimator of s^2.

If the samples have very different means and small standard deviations, it is not necessary to take large numbers of samples. This is also true if the sample means are far from the target level. Only when the level is close to or below the target or desired level will one need to use this equation [6,7].

6.10. COMPARING AREAS

In some cases one might question if all the samples from a field or all the grids in a field are the same or different. To answer this question one might think to carry out a t-test comparing each grid with each other grid. This would be a tedious calculation. Rather than do this, an analysis of variance (ANOVA) can be carried out. To do this the mean or $\bar{x}$ for all the areas or grids in the field is calculated. An $\bar{x}$ for the two areas to be compared is also calculated,* then as with previous calculations a hypothesis is needed. The null hypothesis H_0 is that all the areas are the same.

The two areas represented by B3 through B6 and C3 through C6 in Figure 6.1 can be compared. The null hypothesis would be that the two areas are the same; that is, $H_0 = \mu_{B3-B6} = \mu_{C3-C6}$. An alternate hypothesis H_a might be that the means are different from each other.

The second step is to calculate what is called the sum of the squares for treatment (SST). In this case we assume that grids B3 through B6 are one treatment and grids C3 through C6 are one treatment and that they are the same. The $\bar{x}$ for all measurements is 600, while the $\bar{x}$ for area B3 through B6 is 796 and the $\bar{x}$ for area C3 through C6 is 403. The averages for these areas have been given as three digits without decimal for simplicity. The general equation for calculating SST would be

$$\text{SST} = \sum_{i=1}^{p} n_i(\bar{x}_i - \bar{x})^2$$

In this equation, X_i is the mean of the ith group (i.e., B3 through B6 would be a group), $\bar{x}$ is the overall mean, and n_i is the number of observations in the ith group. (This would be 16 for the B3 through B6 group.) The answer is calculated as follows:

$$= 16(796 - 600)^2 + 16(403 - 600)^2 = 1235600$$

The next step is to calculate what is called the sum of squares of error (SSE). This is based on the variability around the sample means (i.e., 796 and 403). The general equation for this calculation is:†

$$\text{SSE} = \sum_{j=1}^{n_1} (x_{1j} - \bar{x}_1)^2 + \sum_{j=1}^{n_2} (x_{2j} - \bar{x}_2)^2$$

For area B3 through B6 the averages are 1090, 908, 724, and 463, and for area C3 through C6 they are 871, 586, 106, and 51. The averages for these

---

* All numbers have been truncated to whole numbers for illustration. Decimals are kept when working with actual data.

† In this equation j is used in the same way i is used in previous equations.

areas are 796 and 403, respectively, thus

$$
\begin{aligned}
SSE &= [(1090 - 796)^2 + (908 - 796)^2 + (724 - 796)^2 + (463 - 796)^2] \\
&\quad + [(871 - 403)^2 + (586 - 403)^2 + (106 - 403)^2 + (51 - 403)^2] \\
&= 679679.
\end{aligned}
$$

The next step is to calculate a mean square for treatments (MST). This is done by dividing the SST by the degrees of freedom. In this case there are two treatments (p is the number of treatments and n the number of observations, which is 8), and so the degree of freedom is $2 - 1$, which is 1, thus

$$
MST = \frac{SST}{p - 1} = \frac{1235600}{2 - 1} = 1235600
$$

Next is the calculation of the mean square for error (MSE).

$$
MSE = \frac{SSE}{n - p} = \frac{679679}{8 - 2} = \frac{679679}{6} = 113279.8
$$

The last step is to calculate an F statistic.

$$
F = \frac{MST}{MSE} = \frac{1235600}{113279.8} = 10.90
$$

At this point an F statistic shown in Table 6.7 that represents a 95% confidence interval for H_0 is consulted. The degree of freedom for the numerator (v_1) is 1 and the degree of freedom for the denominator (v_2) is 6. Going to the F table the F statistic is 5.99. This could be represented as

$$
F_{.05} = 5.99 \qquad \text{for } v_1 = 1 \text{ and } v_2 = 6
$$

Because the calculated F statistic is more than that found in Table 6.7 it is concluded that there are enough data to say that the two areas are different. This is an important conclusion because it means that the two areas must be treated differently.

This type of calculation can be used to show areas in which the application of a single method of remediation to the whole area as opposed to individual grids is reasonable. It may also allow for reducing the number of samples and consequently the number of analyses that need to be carried out [8].

TABLE 6.7 Abbreviated Table of Critical F Values

$v_2{}^a$	v_1 Numerator degrees of freedom	
	1	2
1	161.4	199.5
2	18.51	19.0
3	10.13	9.55
4	7.71	6.94
5	6.61	5.79
6	5.59	5.14
7	5.59	4.74

[a] Denominator degrees of freedom.

The above calculation using eight means rather than the 32 observations is not a good approach because it does not account for variability in the observations. It assumes that grids B3 through B6 are equivalent and that grids C3 through C6 are equivalent, which is a huge assumption. It is possible to use a more complicated version of ANOVA to test both the differences among grids and between areas at the same time; however, such a procedure is beyond the scope of this book. Two areas could be tested using the t-test; however, it is subject to interpretation errors when applied to more than three areas.

6.11. LINEAR REGRESSION

Linear regression is used extensively in analytical procedures to determine the accuracy of calibration curves. These calculations are thus extremely important because they determine the accuracy of many analytical results. Because these are mostly used in the commercial laboratory they will not be discussed here.

6.12. GEOSTATISTICS

There are many tools that have been and are used for estimating the unknown amount of a component when it is between or surrounded by known amounts of the same component. Simple linear interpolations using nearest neighbors, distance and a weighting factor, and other methods have been developed and used. A method that has gained much use and recognition is called kriging, which is based on geostatistics [9]. In this discussion mathematics will be minimized and the results of various geostatistical operations illustrated without the corresponding calculations.

The basic assumption of geostatistics is that samples taken close together are more similar than those taken further apart; that is, there is a definable relationship between two components. Another way of thinking about this is that one component is correlated to the value of another component. A further assumption is that the changes occurring in the environment occur in a regular, describable fashion over definable distances. The distances used can vary, depending on the component and the needs of the analysis. Technically distances can be anywhere between the atomic and the km scale. The distance over which this influence is active can be estimated and used in calculating the value of the missing data.

For field sampling these assumptions are put into mathematical form so that estimations can be made at unsampled locations. Note that geostatistics is closely related to modeling, which will be discussed in detail in Chapter 7. Indeed, geostatistical methodology and calculations are essential in many environmental models.

Determining the distance over which measured values are interdependent is essential if these calculations are to be useful. Some reported values of the range of interdependence for pH measurements in two locations are given in Table 6.8. As can be seen from these data, interdependence can sometimes range for large distances [10–12]. Interdependence can also be scale-dependent in that it can be different, depending on the overall area of interest.

Kriging could be used to estimate the missing values in A2 and A6 in Figure 6.1. In this case the four nearest values could be used in the calculation; that is, for the missing value X in A2 the value could be estimated using 190, 210, 510, and 321, and the missing value Y in A6 could be estimated using 164, 94, 106, and 500. For this calculation the distance between all points must also be known.

6.12.1. Correlograms

The first step is to think about how environmental measurements might change over a distance. If there is a point source of contamination, we might expect that the level will be high at the source and decrease with distance from the source. In Figure 6.1 the point of contamination would appear to be area B3. A graph of the averages for each of the areas from B1 to B6 is shown in Figure 6.4. The TPH values are seen to decrease with distances from the source. This is true no matter which direction is taken away from area B3. A regular decrease in the component over distance is thus one possible change.

TABLE 6.8 Ranges of pH Interdependence Reported for Two Locations

Location	Soil	Range (m)	Conditions	Source
Arizona	Pima clay loam	1.5	Four 20-meter transects, 20-cm spaces, 50-cm depth	Gajem [10]
	Pima clay loam	21	Four 20-meter transects, 2-meter spacing, 50-cm depth	Gajem [10]
	Pima clay loam	260	One transect, 20-meter spacing, 100 points, 50-cm depth	Gajem [10]
Hawaii	—	14,000–32,000	Transects on island of Hawaii, 1-to-2-km spacing, 0–15 cm deep	Yost et al. [11]

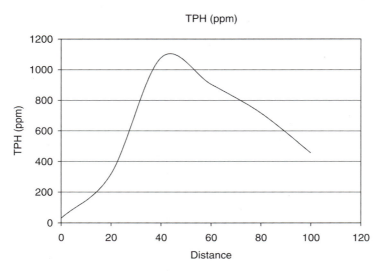

FIGURE 6.4   Decrease in TPH concentration from a point source.

Several other possibilities can exist; however, only two will be discussed here. One would be that the concentration or characteristic does not change over the distance of a transect or sampling blocks. For instance, it might be possible that within experimental error the pH of a soil does not change over the entire area being sampled. In this case the characteristic would follow a model with a slope of 0; that is, there is no change in the value with a change in either the x or the y direction. In an absolute or realistic sense this would not happen. What actually happens is that the pH varies around some mean and the means at all places are not significantly different from each other.

The second would be that the characteristic changes in a regular fashion over the distance of the transect. In Figure 6.5 soil bulk density changes in a regular pattern because of wheeled traffic across the field. This type of graph may result from measurement of any environmental characteristic or quantity, whether it is chemical or physical.

In the types above, if the expected values are dependent on position, then universal kriging, which uses variograms (see below) is used. It might also be noted that at a finite variance both correlograms and variograms contain the same information.

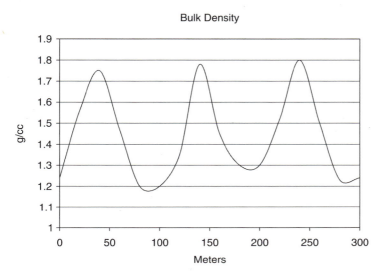

FIGURE 6.5 Recurrent pattern of soil bulk densities due to wheeled traffic across a field.

6.12.2. Variograms

A variogram is a graphical representation of the variation in a measurement over distance and is a basic component of any kriging. Four different types of variograms plotted as variance versus distance are used as models.* (See Figures 6.6–6.9.) Except for Figure 6.8 (the linear model that does not have a corresponding covariance function), each of these can have a covariance function, and this can be plotted and also used in calculations. In Figure 6.6 the variance is 0 at a distance of 0. It increases linearly in a straight line to a value at which it becomes a flat line, a slope of 0. The y value where the line has a slope of 0 is called the sill. The distance from the start, $x = 0$, $y = 0$, to where the slope becomes 0 or where the y axis reaches its maximum value, is called the range. The range is the maximum distance at which components are still correlated. The graph in Figure 6.6 is called a linear model with a sill.

A second linear model (Figure 6.7) has the variance at some value above 0 at 0 distance. The variance increases linearly to some value at which

---

*Although the term modeling is applied to these graphs they are not the same as the modeling we will discuss in Chapter 7.

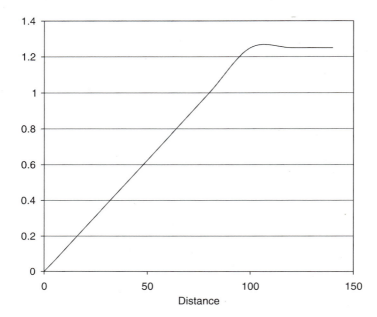

FIGURE 6.6   Linear model with sill and range.

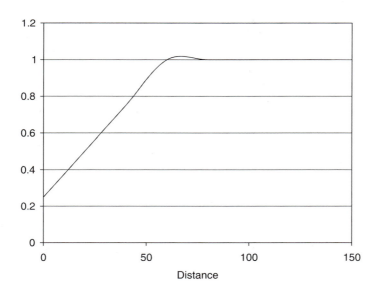

FIGURE 6.7   Linear model with nugget, sill, and range.

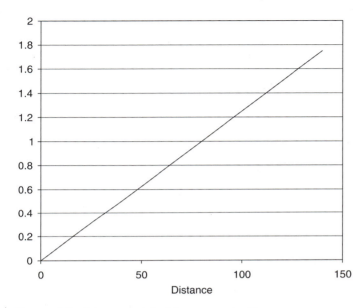

FIGURE 6.8 Linear model without nugget, sill, or range.

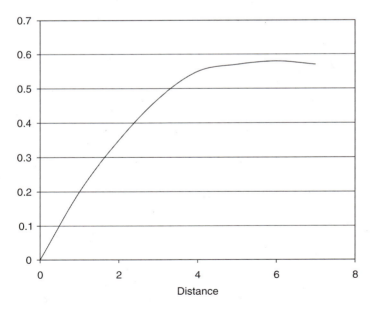

FIGURE 6.9 Spherical model with range and sill.

the line becomes flat with a slope of 0 or the y axis has a maximum value. The variance at 0 distance is called the nugget. The distance from 0 to where the slope becomes 0 is called the range. In this case the variance at the point at which the line has a slope of 0 is made up of the nugget and an additional variance, which is still called the sill. It is important to remember that many of the components being sampled for in the environment occur naturally at a low level. This natural low level and its variance will lead to the occurrence of a nugget in the variogram of this component.

The third model is the linear model (Figure 6.8), which is simply a straight line with no sill. A fourth model (Figure 6.9) is one that is curved between the 0 variance and 0 distance to a distance at which the line again has a slope of 0. The range is from 0 to the place at which the slope becomes 0, and this is the sill. Typically one calculates the variance and plots it versus the distance, and the curve is fitted to one of the appropriate models.

When using data from field sampling the component variogram may not look like any of those shown in Figures 6.6 through 6.9. There are a number of reasons for this. If the concentration of the component goes to 0, there will be no variation, and so the variance will also go to 0. This would be the case where there is a spill of a component that is not commonly found in the environment. At some distance the concentration will be 0 and so will the variance. In this case the variance will increase for some distance and at some further distance fall back to 0. There are other situations that produce variograms significantly different from those shown. In all these cases the graphed variogram is compared to these idealized variograms. The one it fits best is the one used in carrying out calculations.

To carry out a calculation using kriging the distance from the unknown point to all of the known points must be known. The distances between all of the known points are also needed, and thus must be calculated. Two tables are produced, one of which contains the coordinates of each point and a table of distances between all known points. These distances are used in the calculation of both a covariance function and a variogram, and are the basis of the final weighting factors. The variogram is then fitted to a model variogram.

Commercially available software packages are available that fit variograms to a model variogram semiautomatically. However it is important for the researcher to know what is being done, because a number of interpolation methods will be available to choose from. Kriging can be chosen as one of the interpolation methods. (See Chapter 7 on modeling and Appendix A for software sources.)

Kriging methods are used extensively in producing surface and subsurface maps of contamination, thus this methodology will be associated with a wide variety of commercial modeling programs. For these programs actual calculation is not necessary. It will be beneficial to know something about the methods in interpreting the results of this modeling, however. Some understanding of kriging will also help in understanding the steps a commercial program is asking you to carry out.

6.12.3. Kriging Map

The values for a characteristic can be plotted on a map of the area sampled. The maps look like contour maps except they show lines of equal concentration or equal values of some physical constant or chemical characteristic rather than elevations. The lines can be found by using known values and using kriging to calculate the missing or unknown values between the known values. A similar map can be drawn of the variance over the same area, if desired. The type of map obtained is shown in Figure 6.10.

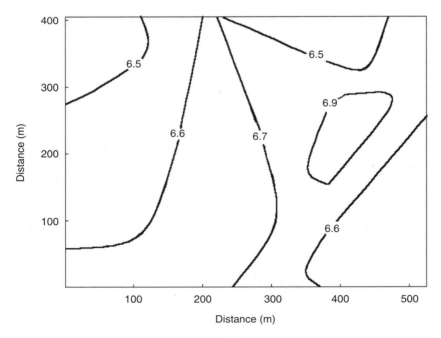

FIGURE 6.10 A map of kriged values. Both a point-kriged and block-kriged map would look similar.

6.12.4. Verification of Kriging Results

An interesting characteristic of kriging is that it is possible to check calculations against values that are later collected or from a group of values not used in the original estimations. A missing value can be calculated using kriging, a sample can subsequently be taken from the field at the missing site, and the analytical results compared to the calculated results. If the geostatistical calculations are valid, then the calculated and new measurements are reasonably close in value.

Another approach would be to use surrounding values to calculate a known value. At this point we would assume a known point is missing and calculate its estimated value from the previously developed equation. The two results should be found to be very close in value. This method of checking the values is called "jackknifing."

6.12.5. Block Kriging

Analytical results from point sources are used in calculating kriging estimates; however, the values from an area can also be averaged and these values used to produce a map of that characteristic. For example, the data in the sample areas in Figure 6.1 can be averaged, as shown in Table 6.7, and this average can be used in block kriging. A map produced by block kriging is generally smoother than the map made from point values. (See Figure 6.10.)

6.13. COREGIONALIZATION

Sometimes two different properties are related to each other. For example, plant wilting point and soil organic matter are closely related. When such a relationship exists, a procedure called cokriging can be carried out. When two characteristics are closely related additional information for predicting one characteristic based on measurements of the second is possible. In some situations this is exactly what is done in modeling; that is, a component related to or dependent on already known characteristics is used to estimate the concentration and movement of the unknown component.

The whole area of geostatistics is mathematically intense, involving a great number of variables and mathematical operations. The calculations themselves are not difficult, but they do involve discrete mathematics and calculus. A number of computer programs are available to carry out geostatistical calculations and can be purchased at reasonable cost from firms such as RockWare. (Information for contacting this company is given in Appendix B.)

6.14. GEOSTATISTICAL PROBLEMS

It would be nice if applying geostatistical methods could solve all environmental problems. Unfortunately this is not the case. Geostatistics works best in areas that are relatively homogeneous and the variables well known or studied. Applying kriging to a field without knowing anything about it can be expected to lead to inaccurate estimations. A 100-ha field uniformly made up of a sandy soil so that all the horizons and their characteristics are the same or are well defined will be amenable to kriging. If the field contains a compacted area, however, this area may not be amenable to kriging applied to a transect line trough that area. This is particularly true in two instances.

The first instance would be where a compacted area smaller than the range is encountered. If this range is used to identify sampling intervals, the compacted area may be missed. In this case two problems can develop. First, if the compacted area is missed its effects on the component of interest will not be taken into consideration in making a management decision, such as a remediation plan. Second, the area on one side of the compacted area may react very differently to the component of interest, and this difference may be missed.

The second instance occurs where the subsurface changes dramatically, although the soil surface appears to be the same. If a sandy, textured soil is bordered by a field with boulders and rocks just below the surface, the characteristics of that area will not be the same as the field to which kriging is being applied. Application to this rocky area will give inaccurate results.

The above situations should not present a problem if sufficient samples are taken and used in the kriging estimation. It is essential that there be a number of component data points representing the entire area of interest [12–15].

I know of one example in which such a situation resulted in a catastrophic dam collapse. A dam was built on an underlying rock layer. Unfortunately there was an area of unconsolidated rock at a distance less than the range that was used in sampling, and thus was not detected. This eventually led to undercutting of the dam and its collapse. This resulted in massive flooding downstream.

6.15. CONCLUSIONS

Statistics is an essential tool in sampling. It is used to determine if two components have the same or different levels of a component of interest. It is used to determine values of interest at unsampled positions. It is also used to estimate the number and positions of samples needed. Geostatistics is useful in estimating contours and missing values from data sets. Various geostatistical methods are used in commercial modeling programs and so are valuable in the modeling of environmental situations. It is highly recommended that a statistician be consulted in using statistics in a sampling plan. It is also highly recommended that appropriate computer statistical programs be used in carrying out all statistical analyses.

QUESTIONS

1. Using a hand calculator or a computer generate 10 random numbers.
2. List and explain the four types of hypotheses.
3. Calculate the median and mean for the TPH values given for areas C1 through C6 in Figure 6.1.
4. Calculate the standard variation of the TPH values given in sampling areas A1 through A6 in Figure 6.1.
5. Using the t statistic determine if the areas C1 and A6 represent the same or different populations.
6. Determine if the value 346 ppm for TPH in sample area B2 is an outlier or is from the same population for the area.
7. The transect sampling of a field shows that the high TPH value is 2000 ppm. The cutoff for TPH in this state is 500 ppm. How many samples will be needed to be sure that all areas in which the TPH is above the allowed level are accounted for?
8. Describe the basic assumption of geostatistics.
9. Explain the terms nugget, range, and sill as applied to kriging.
10. What types of graphs illustrate the types of variations that can be found in analytical results from samples taken over a distance?
11. In kriging, what are four models used to evaluate the component variation?
12. Describe some of the potential problems associated with kriging.

REFERENCES

1. Kempthorne O, Allmaras RR. Errors and variability of observations. In: Klute A, ed. Methods of Soil Analysis: Part 1—Physical and Mineralogical Methods. 2nd ed. Madison, WI: American Society of Agronomy and Soil Science Society of America, 1986:1–31.
2. Freund JE, Simon GA. Statistics: A First Course. Englewood Cliffs, NJ: Prentice Hall, 1991:316–322.
3. McClave JT, Benson PG. A First Course in Business Statistics. 4th ed. San Francisco: Dellen, 1989:80–83.

4. Freund, JE, Simon GA. Statistics: A First Course. Englewood Cliffs, NJ: Prentice Hall, 1991:340–349.
5. Dixon WJ, Extraneous values. In: Klute A, ed. Methods of Soil Analysis: Part 1—Physical and Mineralogical Methods. 2nd ed. Madison: American Society of Agronomy and Soil Science Society of America, 1986:80–90.
6. Crépin J, Johnson RL. Soil sampling for environmental assessment. In: Carter MR, ed. Soil Sampling and Methods of Analysis. Ann Arbor, MI: Lewis, 1993:5–18.
7. Freese F. Elementary Forest Sampling. Agriculture Handbook no. 232. U.S. Department of Agriculture, Corvallis, OR: Oregon State University, 1962.
8. Freund JE, Simon GA. Statistics: A First Course. Englewood Cliffs, NJ: Prentice Hall, 1991:356–365.
9. Myers DE. What Is Geostatistics? http://www.u.arizona.edu/~donaldm/homepage/whatis.html.
10. Gajem YM. Spatial Structure of Physical Properties of a Typic Torrifluvent. M.S. thesis, University of Arizona, Tucson, 1980.
11. Yost RS, Uehara G, Fox RL. Geostatistical analysis of soil chemical properties of large land areas. I. Semi-Variograms. Soil Sci Soc Am J 1982; 46:1028–1032.
12. Warwick AW, Myers DE, Nielson DR. Geostatistical methods applied to soil science. In: Klute A, ed. Analysis: Part 1—Physical and Mineralogical Methods. 2nd ed. Madison, WI: American Society of Agronomy and Soil Science Society of America, 1994:53–82.
13. Sharov A. 2.4 Elements of geostatistics. http://www.gypsymoth.ento.vt.edu/~sharov/PopEcol/lec2/geostat.html.
14. Rossi RE, Mulla DJ, Journel AG, Franz EH. Geostatistical tools for modeling and interpreting ecological spatial dependence. Ecol Monogr 1992; 62:277–314.
15. Isaaks EH, Srivastava RM. Applied Geostatistics. New York: Oxford University Press, 1989.

7

Modeling

Many types of models can be made of the environment; for example, a static physical model, showing the different parts in three dimensions, or a similar dynamic model, which actually models changes in the environment when conditions change. Another approach is to develop a mathematical model that models the movement or actions and reactions of a single component or multiple components in the environment. The ideal would be to have this model be as simple as possible.

Today most mathematical models are developed using computer programs. The advantage of using computers is that models incorporating many variables and complex calculations can be developed and will run quickly and easily on a computer. Mathematical models can be three-dimensional and be presented as relatively simple three-dimensional graphs. On the other hand, the output may be a three-dimensional diagram of the environment showing the surface and subsurface location of the components of interest. These can also be dynamic models and can be used to predict and quantify changes in the environment when conditions change.

All models have two basic valuable characteristics. The first is to allow the visualization of the situation as it exists. Seeing the situation in even a static model can help in the development or modification of a sampling plan. The second is that the model allows the prediction or estimation of changes in the future. Dynamic models allow for changes, which make it possible to predict how these changes would change the dynamics of that portion of the environment being modeled. For instance, a model of soil moisture levels might be developed for a certain rainfall, perhaps 500 mm of rain spread over 6 months. The model would allow prediction of the changes that would be expected if, for example, rainfall was 1000 mm or 250 mm over the same length of time.

Such changes in rainfall will affect sampling. A soluble pollutant will move further and deeper under heavier rainfall conditions and thus require more sampling and deeper sampling. Under reduced rainfall less movement would occur, and thus less sampling is needed. The predictive nature of a model can thus be of great benefit in designing a field sampling plan. In conjunction with sampling, such models can be used in decision making, such as whether or not a contaminated groundwater plume requires cleanup. Such models can guide the location of sampling sites and the drilling of monitoring wells. Models should never be used as the sole basis for location of sampling sites and wells in a field sampling plan, however.

Many constants are needed for any modeling. Some are chemical constants and others are physical constants. A list of common constants and variables used in environmental modeling are given in Table 7.1.

TABLE 7.1 Common Constants and Variables Used in Modeling the Environment

Quantity/constant	Units	Common abbreviation
Water solubility	g/ml	C
Diffusion coefficient	cm^2/s	D
Soil hydraulic conductivity	cm/s	θ
Solar radiation	$erg\,cm^{-2}s^{-1}$	R
Temperature	°C or °K	C or K
Heat capacity	$erg\,g^{-1}d^{-1}$	c
Acceleration due to gravity	m/s^2	g
Soil porosity	cm^3/cm^3	n
Volumetric water content	cm^3/cm^3	θ
Soil matrix potential	kPa	Ψ_m
Saturated soil water content	cm^3/cm^3	θ_s
Volumetric air content	cm^3/cm^3	Φ
Air density	g/cm^3	ρ
Soil bulk density	Mg/m^3	B.D.

Note: Not an exhaustive list.

7.1. PHYSICAL MODELS

A physical model may be constructed in a number of different ways and may represent one or several characteristics of the environment. The model may be static and only be a scale model of the actual area, or it might be dynamic but qualitative in that several components interact and the interactions are observed. In this case the model is used only for visualization. The most sophisticated models will be both dynamic and quantitative, allowing the calculation of the amount of change taking place over time. All of these models may be valuable, depending on the system and the situation being investigated.

7.1.1. Static

The static model typically has all its components represented by various constructed structures. Soil might be represented by fine sand. The sand might be colored to represent various horizons. Fractured rock can be represented by gravel and consolidated rock by colored clay. Water could be represented by water, colored clay, and the like. Such a representation might also show caves, trees, and so forth. Often such a model is constructed to scale. Such a scale might be something like 1:10,000 meaning 1 cm in the model represent 10,000 cm in the real landscape. Using a scale model allows one to calculate distances between features.

Such a scale model can be used to locate sample sites and monitoring wells. In addition, underground features (e.g., pipelines, caves, buried tanks) can be represented in such a way that samples can be taken so as not to disturb or interact with them. Note that in a scale model the position (north, south, east, west) and depth of the features must be exactly known for their correct placement in the model and subsequent avoidance.

Another use of such models would be for comparisons before and after. In this case two or more models can be constructed showing the present situation and the situation as it is envisioned during the project and when the project is completed. This will allow the involved parties to envision the changes so that they will not be surprised by proposed changes or by their effect on the environment.

7.1.2. Dynamic

A dynamic model may be constructed in a manner similar to a static model. In this case, however, the model is interactive in that it shows the interaction between components in the environment. For the environment this type of model often involves observing the interaction between water and soil. It might also show the interaction between air and soil or all three, however.

Many qualitative models of the interaction of soil and water are available. In Chapter 2 there is a model that illustrates the movement of water through soils with horizons having different textures. Another model is to drop water on soil peds held on a wire screen. The stability of the peds can be seen by how they hold up to running water. If water is allowed to drop from some height onto soil it will be splashed. The drops dislodge soil particles that are then carried out of the soil for some distance. Water falling just a few meters can cause soil to splatter a meter or more from the point of impact. (See landscape modeling in Section 7.4.)

If a board with ridges on three sides is constructed and filled with soil, it can be used to model stream movement and development. (See Figure 7.1.) To protect the wood from water, the board is often covered with plastic before the soil from the area being investigated is added. The board is placed at an angle to represent the slope of the area being investigated. A source of water is allowed to flow across the soil. As the water moves across the soil it will form a "stream" that mimics the actual movement and action of natural streams. In this way the possible movement of contamination in a field may be estimated before the field is sampled. It will also show where field offices and storage facilities are best located.

This type of model is also valuable in observing the effects of flooding on an area. The amount of water flowing across an area can be doubled, tripled, or made 10 times larger, and the effect of these changes noted. If the area is particularly prone to periodic flooding this type of model can be invaluable in locating both sampling sites and sites for buildings. This type of model is shown in Figure 7.1.

Blowing air across soil or sand when it is wet or dry will model what happens during wind erosion. Such a model can be used to design barriers or other structures to prevent or lessen wind erosion. This can thus help in understanding how to design a sampling program during which wind could move pollution onto or off a contaminated field.

In developing and using models it is important to be aware of scaling. Some characteristics may be different or act differently at different scales. What works at a 1-meter scale may not work at a 100-meter scale. Sometimes it is possible to use nondimensional representations so that the values are the same at all scales. Sometimes the scaling problems are not readily obvious. They may even be hidden, so that they only become apparent when the scale is applied to a larger or smaller area. This can be easily demonstrated by filling a jar or glass full to the rim with water, placing a #10 sieve over the mouth, and quickly flipping it upside down. The water will flow out. Repeat the procedure using a #200 sieve, and it will be seen that the water does not flow out. Predicting the movement of water in small pores from models using large pores will fail.

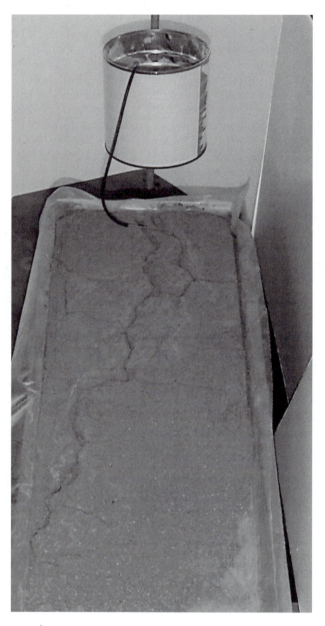

FIGURE 7.1 Modeling stream formation and development. Note the development of the stream channel and its path across the landscape.

7.1.3. Quantitative

In many instances quantitative rather than simple qualitative models may be desirable. Here again several types of physical models are commonly used. The most important and commonly used among these is the soil column. Columns from several centimeters in length and diameter to several meters in length and decimeters in diameter are frequently constructed to model the movement of water and contaminants through soil. Glass, plastic, and metal have been used to contain the column, with plastic being the material of choice today. Such columns are often fitted with ports, electrodes, and other monitoring devices so that the conditions in the column and the movement of components through the column can be followed.

Columns can be constructed in a number of different ways. A retaining material, screen covered with vermiculite, or coarse sand covers the bottom. This is often constructed so that the flow out of the column can be controlled. Air-dried soil is added to the column and then water is used to saturate it. Alternatively the column can be partially filled with water, and soil can be slowly added. The latter method produces a more uniform soil column, although it may be less representative of soil in the environment. Water, contaminants, or solutions of contaminants in water can be added to the columns and movement through the column followed. A column used for modeling water and contaminant through soil is shown in Figure 7.2.

Soil columns constructed and used in this way can be very useful in determining the fate of contaminants, insecticides, herbicides, and so on in soil. A graph of the movement of ethers dissolved in water through the soil column shown in Figure 7.2 is shown in Figure 7.3. The units of the x axis are apparent void volumes. The apparent void volume is void space in soil. (See Chapter 2.) The more soluble methyl tertiary butyl ether (MTBE) elutes from the column before the less soluble octyl ether. Note that the more soluble compound comes out a little before one complete void volume. This is attributed to preferential movement through large pores in the soil.

The data obtained from the graph in Figure 7.3 can be used to predict where a contaminant is most likely to be found in a soil profile and how far and fast it will move, and can be estimated on the basis of the mm of rainfall. From these data the breadth and depth of needed sampling can be estimated. It should be noted that such column conductivities are often lower than field conductivities. This can be due to plant cover, soil structure, cracks, and animal holes. When needed, field studies using lysimeters can be used to validate column studies.

Lysimeters are another dynamic physical way of modeling the environment. A lysimeter is constructed by isolating an undisturbed volume of soil in the field. The soil is isolated by concrete or metal dividers set in the

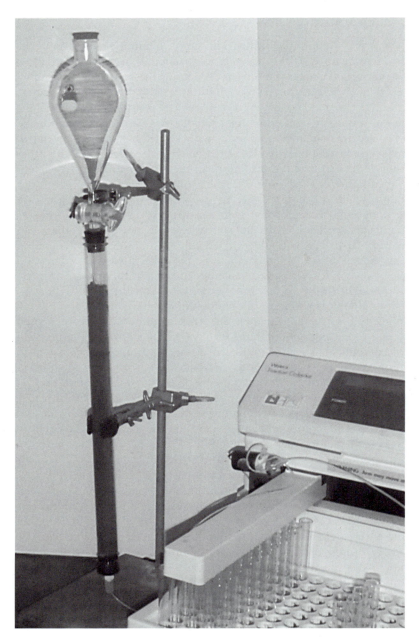

FIGURE 7.2   Soil column for modeling water and solutes movement through a soil.

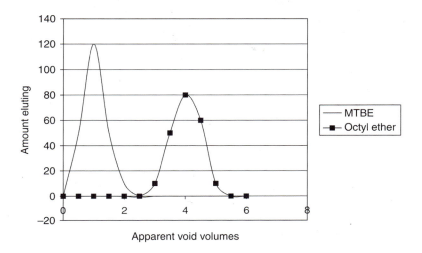

FIGURE 7.3 An example of the movement of ethers through the column shown in Figure 7.2. The y axis is in arbitrary units. The x axis is in column apparent void volumes. MTBE is methyl tertiary butyl ether. (Data from senior research project of Dan Tener, Chemistry Department, Wilmington College, Wilmington, Ohio.)

field. The bottom of the soil is retained by a metal grill supported by sand and gravel underlain by a slanting, water-impervious layer of plastic, cement, or metal. This allows water percolating through the soil to be collected. The big advantage of lysimeters is that they contain a largely undisturbed column of soil. Because it is undisturbed it is expected to more faithfully model the action and interactions of the soil as a whole. A diagram of a typical arrangement of a lysimeter is shown in Figure 7.4.

Water collected at the bottom of the lysimeter can be pumped to the surface for sampling and analysis. Alternately, an access tunnel can be constructed to allow samples to be collected without pumping. There are too many different designs of lysimeters to describe them all here. In some cases a soil water sampling methodology that collects soil water at some depth may be called a lysimeter even when it does not involve a physical lysimeter as such.

Movement of water and contaminants through a lysimeter can be followed under natural rainfall or irrigation. The amount of water percolating through a soil versus the amount lost through evapotranspiration can be measured and correlated to environmental conditions. Knowing the amount of percolating water is essential when modeling soil phenomena.

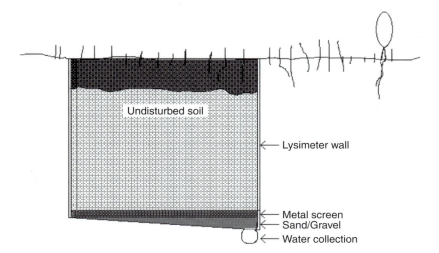

FIGURE 7.4   A lysimeter containing an undisturbed soil and a water collection sump.

7.1.4. Evapotranspiration

In the field, especially in a field with plants, measuring and separating evaporation and transpiration is extremely difficult. Evaporation is the loss of water from the bare soil surface. The soil surface is under continually varying conditions of cover, shade, direct sunlight, and so on. Transpiration is the loss of water from stems and leaves. Stems and leaves lose water at various rates, depending on air temperature, relative humidity, sunlight intensity, wilting point, stage of plant growth, and so forth. At a minimum there are six interdependent, constantly changing variables that must be measured to predict water loss. It is much easier and more accurate to measure the total water loss during the growth of the plants, called evapotranspiration, and even this is generally very complicated.

To simplify estimating or modeling the loss of water during a period of time an open or evaporation pan method is usually used. A pan filled with water is placed in the field and the amount of water lost each day is measured. This is then related to environmental conditions and water loss from evapotranspiration using a pan coefficient. The concept is that the evaporation pan integrates the effects of temperature, wind, relative humidity, solar radiation, and so on, and so can be used to estimate evapotranspiration. Today other more sophisticated instrumentation is becoming more widely used.

The pan coefficient would seem to be unrelated to models needed or used in field sampling. With volatile contaminants, however, loss from the surface may be significant. Although it is often assumed that contaminants can only move deeper in soil, this is not true. When evapotranspiration exceeds precipitation, components can move up the soil profile. Evapotranspiration data can be used to predict where components will be in a soil or regolith profile. This in turn will be important in designing a sampling plan.

Also, the data obtained from evapotranspiration studies will be needed for some mathematical models to be discussed below [1].

7.2. CHEMICAL MODELS

Chemical-type models of the environment usually rest or depend on some underlying chemical principle. Kinetics, equilibrium, solubility, mass transfer, and partition between organic and aqueous phases have all been used in modeling compounds in the environment. These approaches work best with single chemical species or two or more similar species that do not interact with each other. Many mathematical models are also based on chemical phenomena.

Thermodynamics is the study of energy (heat), its movement, and its effect on its surroundings. All systems generally move from conditions of higher energy to conditions of lower energy. They also move from conditions of higher organization to areas of lower organization. In thermodynamic terms this would be moving to conditions of greater randomness. In the environment water moves downhill. Water also moves down through soil and is held more strongly in small pores and less strongly in large pores. To move it from small to large pores requires the input of energy. Organic matter in aerobic soil is oxidized to carbon dioxide and water. In anaerobic soil organic matter is also broken down; however, in this case the end product is methane. In both these cases energy is given off and the compounds produced are smaller and more disorganized [2]. If a reaction results in energy being given off and randomness increasing, then it can be expected to take place.

Looking at the large scale, the soil is a medium designed to break down organic compounds. It provides an intimate mix of solid surface, inorganic nutrients, water, oxygen, and living organisms, and all of this is in the correct proportions to promote rapid decomposition with the consequential release of energy and increase in randomness. Using all of the above the fate of a component in the environment can be modeled by knowing and following changes in its energy. The component will move from situations of higher energy and more organization to situations of lower energy and less organization.

Equilibrium is also used to develop models of the fate of components in the environment. The concept of equilibrium is that at a macro scale a condition is reached at which the concentrations of components on either side of a chemical equation do not change. This situation is most often a dynamic equilibrium, however, in which reactants and products are moving back and forth between these two conditions. A standard representation of a chemical equilibrium reaction is shown and is said to be at equilibrium when the rate k_1 is balanced by the rate k_{-1}.

$$A + B \underset{k_{-1}}{\overset{k_1}{\rightleftharpoons}} C + D$$

Equilibrium models such as the one given below are fairly simple and easy to calculate. In this equation the components are assumed to be at equilibrium. The total amount of the component can be found by summing the amounts in the solid, liquid, and gaseous phases.

$$m_{i(\text{total})} = \sum m_{i(\text{solid})} + \sum m_{i(\text{liquid})} + \sum m_{i(\text{gas})}$$

Although over short periods of time the environment may appear to be in equilibrium, it is never truly in equilibrium. Such equations may provide important information about a component in the environment, however. They may also be included in more comprehensive models of the environment [3].

Mass transfer is the movement of a mass of material through a medium. The medium could be air, water, or soil. There are simple chemical equations for mass transfer; however, in soil the situation is complicated because there is a limited amount of space available for transfer. For this reason a capacity coefficient is needed. In soil some volume containing solvent and solute is immobile and so does not contribute to mass transfer. To complicate equations and models is the fact that solute can be sorbed to pore surfaces. If the process is started with dry soil, some solute (water) will be sorbed to the surface so tightly that it will be immobile. This amount of solute is also part of the equation.

Two components of mass transfer are spherical diffusion and advective (horizontal) dispersion. In spherical diffusion a point source of material will tend to diffuse outward from that sphere in all directions equally. Conceptually there may be a number of spheres and material diffusing out of them. Another possibility, advective dispersion, is horizontal movement away from the source. This would be the case in a stratified medium, in which a denser layer underlies a less dense horizon.

The combined soil–atmosphere model for evaluating the rate of loss of surface-applied pesticides developed by Rivka Reichman, Rony Wallach, and Yitzhak Mahrer is an example of this type of model and this type of modeling [4]. This model is specifically designed to be used with the pesticides lindane, dieldrin, and trifluralin. It is developed using thermo-chemical-type mass balance equations for water and contaminant in the atmosphere and soil. Partial differential equations are used for heat, moisture, and chemical transport. Other equations are used for surface and boundary conditions between soil and the atmosphere. Once developed, the model can be run using a computer after input of the appropriate data. The model is then compared to results from an experimental study.

There is a rather large number of inputs used in most models, including the one above. In this particular case the model involves 73 constants, variables, and assumed conditions. In some cases once all the variables are known the answer is also known.

An example of a chemical kinetic model is a modification of the environmental model called GLEAMS (groundwater loading effects of agricultural management systems) [5] to simulate mercury transport. This model is called Hg-GLEAMS. Kinetics are used in this model to represent the transformation of mercury between its various states. These equations are used to measure the rate of conversion of mercury between Hg^0, metallic mercury, and Hg^{2+}. Equations take the form of standard chemical rate equations. This model also requires meteorological, soil water partition coefficients, and kinetic rate constants [6].

Kinetic models are useful when it is important to determine the rate of both production and loss of a single species or component. Ammonia applied to soil is oxidized to nitrite and subsequently to nitrate. The amount of nitrite existing at any one time will depend on both these rates. Assuming a one-time contamination of a field and knowing the removal rate of the contaminant, suitable sampling intervals can be developed that can reduce the amount of sampling needed. If the rate of oxidation of nitrite is known, for instance, the time needed for its level to decrease to acceptable levels or below action levels can be calculated. In this case sampling and analysis can be done on the basis of the predicted rate of loss rather than on an arbitrary schedule and consequently decrease the amount of sampling needed.

Models will also incorporate such factors as vapor pressure and sorption and diffusion characteristics, which are temperature-dependent, and thus the temperature becomes an essential component of the equation. The soil characteristics (i.e., sandy or clayey), along with dry and wet volumetric water contents, are important, along with solar radiation, air temperature, relative humidity, and wind speed. All of these interact to moderate the temperature changes in soil and so they also become part of the equation.

7.2.1. Computer Chemical Models

Computer models typically involve the use of numerical modeling discrete versions of partial differential equations. This is particularly the case for modeling transport processes. Table 7.2 gives a number of common computer programs used to model chemicals in soil. In some cases these are used to model a single species and its form under the conditions specified. In other cases they are intended to be more general. In both cases the models use many assumed or inferred conditions, constants, and inputs. In spite of the limitations many of the models can be useful in understanding the position and movement of a contaminant in soil or in the environment in general.

Model usage involves three steps. The first step is to pick the appropriate model. This may seem simple, but in some cases the suitability

TABLE 7.2   Common Soil Chemistry Models

Model	Source
GEOCHEM	Mattigod SV, Sposito G. Chemical Modeling of Trace Metal in Contaminated Soil Solutions Using Computer Program. GEOCHEM. In: Chemical Modeling in Aqueous Systems. ACS symp. ser. 93. Jenne EA, ed. Washington, DC: American Chemical Society, 1979: 837–856.
SOILCHEM	Sposito G, Groves J. SOILCHEM on the MACINTOSH. In: Chemical Equilibrium and Reaction Models. Loeppert RH, et al., eds SSSA spec. pub. 42. Madison, WI: ASA and SSSA, 1990: 271–288.
MINTEQ	Femy AR, Girivin DC, Jenne EA. MINTEQ: A Computer Program for Calculating Aqueous Geochemical Equilibria. NTIS PB84–157148, EPA-600/3–84–032. Springfield, VA: National Technical Information Service, 1984.
C-SALT	Smith GR, Tanji KK, Burau JJ. C-SALT—A Chemical Equilibrium Model for Multicomponent Solutions. Loeppert RH, et al., eds. SSSA spec. pub. 42. Madison, WI: ASA and SSSA, 1990.
FITEQL	Westall JC, Morel FMM. FITEQL: A General Algorithm for the Determination of Metal Ligand Complex Stability Constants from Experimental Data. tech. note 19. Cambridge, MA: Ralph M. Parson Lab., Dept. Civil Eng., 1977.

Note: ASA, Agronomy Society of America; SSSA, Soil Science Society of America.

of a model may not be apparent until it is in use. The second step is calibration of the model.

The third step is validation, which must be done in all cases; that is, if the model actually predicts what will happen in the environment. Verification is usually done by comparison of a model to a field situation. In restricted areas with uniform or very similar soils models can often accurately predict the form and movement of a contaminant in the environment. Caution is important, however. Often the model will not be as effective or may not predict the field situation at all when it is applied to field situations in other localities; thus for each new application of a model a validation experiment must be done. This is especially true if the model is to be used to predict sampling site locations.

Most computer models contain empirical components that are varied by the researcher to produce an acceptable result. Some models contain several such factors related to various components. Another approach is to use weighting factors. These can be used to increase or decrease the importance of a particular component of the equation so that the results better correspond to what is actually happening in the environment.

All of these types of models suffer from the fact that they need large, sometimes vast, arrays of inputs and variables to make them useful. In many cases these variables and constants are not known. In these cases they can be obtained either by making field measurements (measuring a soil's porosity, bulk density, etc.) or using column and lysimeter studies. Another approach is to estimate by various means the numbers from other known characteristics of the soil or landscape. This is often done during a calibration stage. For example, in groundwater (GW) modeling, well pumping, along with pressure head measurements, might be taken to calibrate a GW model. In this case, calibration is done by, for example, adjusting the value for soil conductivity. Once calibration is completed other data are used for the verification phase. After the verification phase and once all the data are collected or estimated a calculation can be made.

7.3. SIMPLE MATHEMATICAL MODELS

A number of soil samples from differing soil types, specifically different soil orders from different places, can be used in a series of experiments determining different characteristics. When all the data are collected and all the analysis completed a relatively simple multiple regression analysis can be carried out using common statistical computer programs. (See Chapter 6.) The results of such an analysis will relate the various variables and show which are important in predicting a characteristic of a contaminant in soil.

The sorption of chromium (CrIII) was investigated by Stewart et al. [7] in such a manner. The analysis showed that sorption of chromium was related to a soil's pH, inorganic carbon, cation exchange capacity (CEC), and clay content. They were also able to determine that the bioaccessability of CrIII was related to a soil's clay and its inorganic carbon content. Both of these were represented by relatively simple equations called models of CrIII sorption and bioaccessability in soil.

7.4. LANDSCAPE MODELS

One of the most used environmental soils models is the universal soil loss equation (USLE) or the updated revised universal soil loss equation (RUSLE). This model has been developed for the United States; however, it and similar equations are used in other countries. The use of this equation involves two components. The first part is an equation that allows the calculation of the estimated amount of soil lost (eroded) from a field in a year under a set of described conditions. The second part is a t-value for the soil or soils in the field or fields. The T-value is the acceptable loss of soil in tons per year.

The USLE takes the form of

$$A = R \times K \times LS \times C \times P$$

A is the amount of soil lost in tons per year. This figure is compared to the acceptable loss of tons per acre of soil from the soil types present in the field. The acceptable level is the amount that can be lost without adverse affects on the soil or the surroundings. It is important to know that even under the best vegetative cover soil is lost by erosion every year. This is called geologic erosion. Accelerated erosion is above this natural level and is caused by man's activities. The T-values to which A is usually compared range between 1 and 3 tons per ha per year.

The R factor is a storm erosivity factor. It takes into account the intensity of rainfall, which is a measure of the energy in the raindrops and the rate of rainfall. The energy in rainfall depends on the size and speed of the raindrop when it hits the soil surface. The size of the raindrop is the main factor because the speed is determined in large measure by gravity. Basically this means that rains that come in short, heavy downpours, with large drops are more erosive than long, gentle rains. The rate of rainfall is the amount of rain per unit of time. Typically this figure is in cm/min (in the United States in./hr). The higher the rate of rainfall, the more erosive a rainfall event will be.

R factors are measured or interpolated. A common way to represent R factors is as lines of isoerosivity on a map of the area being studied. In the United States such maps are available for selected areas, and the whole country. The lines are separated by 50 R units. The R factor needed is obtained from this map and inserted into the equation.

The third factor in the equation is the erodability factor, or K. Because of differences in infiltration and percolation rates and structural stability, different soils are more or less resistant to erosion. This difference is reflected in the K factor. These factors have been developed from field studies of numerous soils. Erosion by natural rainfall events from field plots with definite area, length, and slope have been measured and related back to soil characteristics. Either a soil will have an experimental K value or its K can be calculated from the characteristics of the soils in the field.

The L factor is the length of the field and S is the slope. These two are combined to give a single factor, LS. The body of the LS table contains the LS factors as a single number. Factors are found by finding the slope on the left, length on the top, and LS factor in the body of the table. (See Refs 8 and 9 for examples of a table.) From such a table a soil with a slope of 4% and a length of 100 meters would have an LS of 0.64. A soil with a slope of 16% and length of 200 meters would have an LS of 5.40.

Both the C and P factors are obtained from a table relating soil cover, organic matter, vegetation, and management. A field in permanent pasture would have a C factor of 0.003, and with a corn–soybean rotation the factor would have a C of 0.53. The P factor is the conservation practice or practices used in the field. Some examples would be grass waterways, terraces, and contour strip cropping. Conservation practices control the movement of water over the field or decrease the LS factor to lower erosion. A field with a slope of 7% that is strip cropped has a P factor of 0.25, while this same field with 33-meter-wide terraces without water outlets has a P factor of 0.6. All these factors can be found on various Internet sites, such as those given in Refs. 8 and 9.

The appropriate factors from the tables are inserted into the equation and multiplied together to obtain the estimated soil loss A. This is then compared to the T factor for the soils in the field. If the A value is too high, management factors LS, C, and P are adjusted so that a suitable soil loss is obtained. Note that the R and K factors cannot be changed [8,9]!

This equation can be used in a different fashion in field sampling. Here the information needed is where the component of interest is most likely to be found. By knowing the rate of erosion, the position and movement of the component along the surface can be estimated. Places can be identified in which the component may be found deeper in the soil profile because of the deposition of eroded soil. Areas in which the component is closer to the surface because of the erosion of overlying soil can also be determined. The likelihood and rate of soil and contaminant being eroded off the field can be calculated. Also, if contaminated material is eroding onto the field the analytical results of field sampling will constantly be changing, making the sampling program futile unless erosion is taken into account.

The R and K factors can be used by themselves to estimate the amount of soil affected by splash erosion. Splashing can lead to contamination of vehicles and buildings. It will also expose sorbed contaminants to the atmosphere. This can lead to evaporation and loss of material [10].

The RUSLE is the revised version of the USLE. This equation can easily be accessed on the Internet. An explanation of the RUSLE is presented at this Internet site, along with a computer program that allows the calculation of the A factor. The program is not Windows-friendly and is unconventional in its use of keys for moving through the program and obtaining the data needed to make a calculation. It has some interesting features, but will take some time to figure out if real use is to be made of it [9].

There is a similar equation for wind erosion. This equation takes the form

$$E = f(\text{ICKLV})$$

It states that the amount of soil eroded by wind E is a function f of I soil erodability, C climate, K soil ridge roughness, L width of field, and V vegetative cover factors. In this case these factors are not simply multiplied together but are calculated using the appropriate equation. This is the wind erosion equation (WEQ), and there is also a revised wind erosion equation (RWEQ). Calculations using WEQ are carried out using a computer program [11,12].

Wind erosion can complicate field sampling by depositing or removing contaminants, thus as with the USLE and RUSLE, the WEQ and RWEQ can be used to estimate where contaminants are being exposed or buried by windblown soil. It is common for contaminants from manufacturing to be carried in air. These contaminants can be deposited on a field by simply falling out of the air or with rain when it occurs. Erection of windbreaks to stop wind erosion will also allow pollutants to fall out of suspension. This continuing source of contaminants can complicate the sampling and analysis of field samples if not taken into account.

7.5. MODELING MOVEMENT OF PARTICLES
THROUGH THE ENVIRONMENT

Calculation or modeling the movement of particles in air may seem relatively simple, but can be complex. There is a complex interaction of wind, field, and particle concentration. There is also the question of the type of movement being considered. Is it by saltation or suspended transport? Add to this the complexity of turbulent flow, and the system becomes complex. In spite of this it is often the case that time in the air and distance moved is relatively easy to determine. The glide path of a particle can be calculated once the characteristics of the particle are known. This involves not only the pull of gravity but also the shape of the particle. A flat particle will have a longer glide path than a round particle and will take longer to reach the ground. Mass also comes into play in that as the particle size increases the time in the air decreases until it reaches a value of 0 with particles larger than sand.

Movement of suspended particles depends on wind speed and direction. High wind speeds will keep particles in suspension for long distances. The direction of the wind will determine the direction of movement of the particles, thus where the particles are deposited will depend on both these factors, which are relatively easy to determine.

In many ways the movement of particles in water is similar to their movement in air. The speed of particles settling will depend on their shape and mass and the flow of the water they are suspended in. One difference is that some particles will be heavier than water and settle; others can be less dense than water and thus float. Some particles will be so small—colloidal in size—that they will remain in suspension and not settle out. Particles that are heavier than water can be modeled in a manner similar to modeling the movement of the same particle in air. Particles that float or that initially float absorb water and then sink can also easily be modeled.

Particles that remain suspended can be modeled, but the modeling is more complex. Some components in water will be in equilibrium with the particles; others will be irreversibly bound to them. In both cases modeling these particles is essential because they carry contaminants.

There is one other common characteristic of air and water in regard to contaminants. Gaseous and soluble contaminants will mix with and be dispersed throughout the air or water in which they are mixed. There are initially concentration gradients, but over time the solute or suspended material will be evenly distributed through the solvent [13].

Movement of components through soil, regolith, decomposed rock, fractured rock, and whole rock is entirely different. In all these cases an agent (air or water) is necessary to carry the component through the media.

The movement will be tortuous, and solutes will interact with the solid media through which the solvent is moving. The interaction may be chemical, such as bonding to surface groups, or it may be physical, such as diffusion into pores or filtering out. In either case the solute is slowed in its movement. Sometimes its chemical composition is changed by interaction with solids or by being attacked by microorganisms in the environment. Modeling this movement is thus very complex and subject to a great deal of variability.

7.6. DYNAMIC MODELS

Dynamic modeling is complex in that it considers all the eventual fates of a component in the environment. This includes the reservoir of the component, the sources that add to it, and the mechanisms by which it is dissipated. If there are any feedback loops or conditions, these will also be included. Simply knowing the components of the system does not make a model dynamic. What does make it dynamic are the rates of all its transformations, including feedback. Knowing the rates of the various transformations, additions, and deletions allows the calculation of the fate and longevity of the component.

Several possible dynamic systems are illustrated in Figures 7.5 through 7.10. These cases show in a general way how components interact with the environment. Figure 7.5 shows a logarithmic increase, which is very common for microbial growth when an easily decomposed compound is added to soil. Figure 7.6 illustrates the increase with a subsequent decrease in a component in the environment. This type of change is common when an easily decomposed compound is added to soil, and microorganisms produce a component toxic to themselves.

Figure 7.7 shows a cyclic change in a component's occurrence in the environment. Many phenomena follow this pattern, particularly plant and animal populations. Figure 7.8 shows a periodic decay that is characteristic of a microbial population decomposing one component and releasing another that can be used by a different population. The second population breaks down this component, releasing yet another component used by another population. This process continues until all the material has been broken down into carbon dioxide and water.

Figures 7.9 and 7.10 show situations that are not sustainable and are not common, although they may appear to be over short periods of time. Figure 7.9 shows a linear increase or decrease in a component. Components, however, cannot be continuously added or subtracted from the environment. Eventually a level will be reached at which dissipation equals

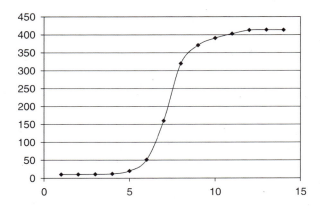

FIGURE 7.5 Logarithmic increase in component in environment.

addition, or else in the case of dissipation the component either is present in very low concentrations or is not present at all.

Figure 7.10 shows a slow increase followed by a logarithmic increase in a component. It also shows a logarithmic decrease in a component followed by a slow decrease. This last situation is common in soil in which some of a component is in solution and easily decomposed while some portion is adsorbed on a surface and some is in small pores that do not drain.

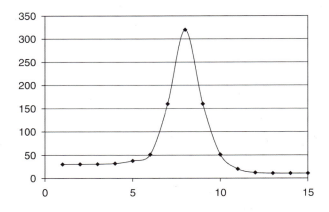

FIGURE 7.6 Increase followed by decrease of component in the environment.

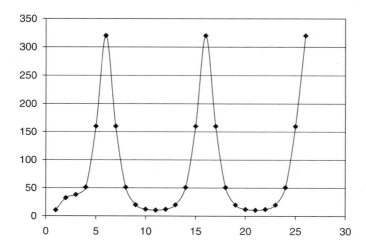

FIGURE 7.7   Cyclic increase and decrease in component in the environment.

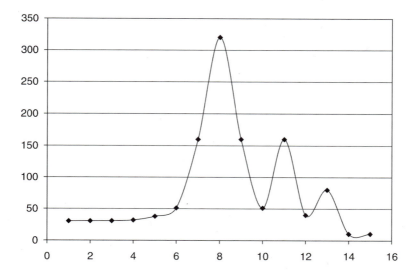

FIGURE 7.8   Periodic decrease in component in the environment.

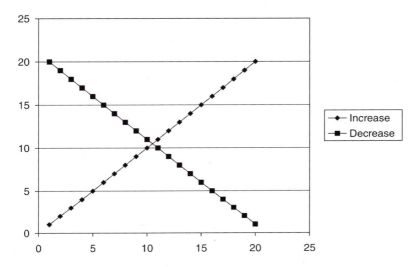

FIGURE 7.9 Linear increase and linear decrease in a component in the environment.

Components adsorbed on particle surfaces will decompose slowly because they are less available when adsorbed. Components in small pores must diffuse out of the pore before they can be decomposed. The rate of diffusion will depend on the concentration difference between the bulk solution and the pore solution. When the concentration of component in the bulk

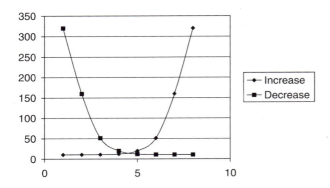

FIGURE 7.10 Exponential increase and decrease of components in the environment.

solution decreases to a point at which it is lower than that in the pore, then the component in the pore will diffuse out and be decomposed. Diffusion is a slow process and therefore the decomposition process will appear to occur at a slow rate. A component's concentration cannot continuously increase or decrease exponentially. Eventually inputs will cease or equal losses or the component will completely disappear from the environment. Sensitive chemical instrumentation indicates that complete loss is not likely.

A concrete example of looking at the changes in a component in soil is soil phosphate. In soil it increases as organic matter decomposes and decreases as plants take it up and as it reacts to form insoluble phosphates. Changes in phosphate in soil could be described by a simple mathematical relationship:

$$\text{Future levels of phosphate} = \text{previous levels of P} + \text{additions of P}$$
$$-\text{losses of P}$$

or

$$P(t + \Delta t) = Pt + Pi - Po$$

where $P(t + \Delta t)$ is the phosphate content at some time in the future (Δt is a change in t), Pt is the phosphate at the present time t, Pi the inputs of phosphate, and Po the outputs or loss of phosphate. This whole relationship can be represented by a simple flow diagram, shown in Figure 7.11. In Figure 7.11 the central rectangle represents the pool of available phosphate

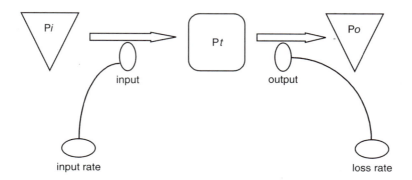

FIGURE 7.11 A simple flow diagram for the input and loss of phosphate from soil (triangles represent Pi (left) and Po (right)).

while the arrows indicate the direction of flow. The circles represent the activity while the oblong circles represent the rate of the activity happening. The rates are connected to the process, indicating that the total flow over time is determined by the rate at which the activity happens.

In developing a dynamic model inputs would be separated into those from fertilizer application, organic matter breakdown, and release from minerals. In a like fashion losses would be separated into plant uptake, erosion, leaching, and precipitation. This would lead to a much more complex but more realistic model. This increased complexity is shown in Figure 7.12 [14].

This type of modeling can be very valuable in developing sampling plans that involve determining natural attenuation, bio- or phytoremediation, or fertilizer use. The rate of loss can be used to predict future levels of the component of interest and thus the need and timing of future sampling.

7.7. COMMERCIAL COMPUTER MODELING

Computers and computer modeling have become essential to the whole process of environmental modeling and field sampling. Computer models

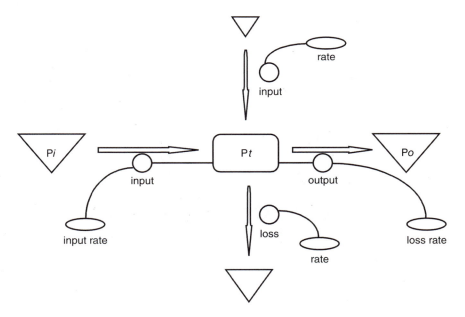

FIGURE 7.12 A complex flow diagram for phosphate in soil (upper triangle might be fertilizer input, lower triangle might be plant uptake).

can make visualization of the landscape or part of the environment possible. Several features make these models extremely valuable. First, they are easy to construct and easy to change when necessary. New information can be added as it becomes available. The new information can then be added to a visual representation of the area of concern. Some programs allow analysis and correlation of data. Such questions as where the contamination ends and how much area is covered by the contamination are questions that the models can answer.

Figures 7.13 and 7.14 show two types of outputs that can be obtained by this type of program. These are representations of the concentration of a pollutant in a field as calculated by a model. Figure 7.13 shows the surface spread of the component and its decrease in concentration as it moves away from the source. Figure 7.14 shows a cross-sectional view of the same contamination as it moves down through a soil profile. The decrease in concentration as it moves away from the source is again shown. Both these situations are relatively easy to model because of the uniformity of the soil and regolith.

An example of a commercial program used to model GW movement is Modflow, which was developed by the United States Geological Survey

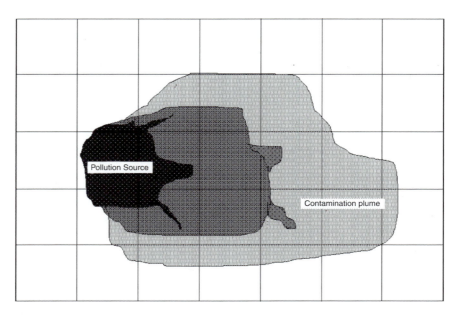

FIGURE 7.13 Surface view of the spread of a contaminant in soil. Lighter areas indicate a lower concentration of the component.

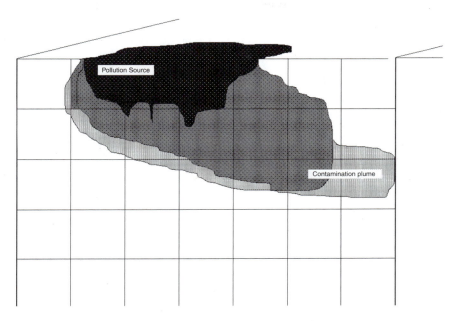

FIGURE 7.14 Model of subsurface movement of contaminant.

(USGS). It used finite-difference algorithms of the governing partial differential equations to model GW flow. There are also contamination transport models that interface with Modflow. (See Table 7.3 and Appendix B for more information about this program.)

TABLE 7.3 Commercial Computer Modeling Programs

Program name/abbreviation	Type of modeling
GMS 3.1	Groundwater modeling
Modflow	Groundwater modeling
SMS	Surface water modeling
ChemFlux	Contaminant transport modeling
ARC ESRI GIS	Data visualization, query analysis, and integration
MrSID Geo	Image file handling[a]

Note: All programs available from RockWare; see Appendix B.
[a] A sample MrSID is available on the Web at no cost.

In Figure 7.15 there is a compacted layer and contaminant is "leaking" through the layer into lower layers. This type of situation is impossible to model if the location of the "holes" in the compacted layer is not known. Even when the holes and their positions are know they are hard to model. The figures shown are black and white; however, most if not all computer modeling software will present these types of results in color. Different colors will be used to show the concentration of the components of interest in various locations. A color scale with reference will also be provided. Some common commercial computer modeling programs are given in Table 7.3.

In developing a field sampling plan these models can help in directing sampling efforts to areas with the greatest likelihood of being contaminated. Another way to use these is in presentations designed to justify sampling plans and procedures to regulatory agencies. The data can also be presented in a way that local populations can easily understand.

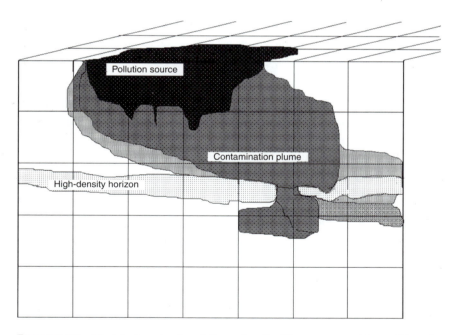

FIGURE 7.15 Model of contaminant through soil with a compacted layer. Note that the contamination has moved through a "hole" in the compacted layer and is contaminating lower layers.

7.8. GIS AND ENVIRONMENTAL MODELING

Because the Geographic Information System (GIS) is a powerful tool for relating and displaying areas on and in the Earth, it has been proposed that it can be used in environmental modeling. Some developments along these lines have been made, but GIS is still used to display the environment as it is presently. It thus does not have predictive capabilities, which are an important component of environmental modeling. By displaying environmental conditions over a period of time, however, trends in the movement of pollution or changes in the quality of the environment can be seen and predictions as to future changes can be made [15].

7.9. WHOLE PLANET MODEL

Koziol and Pudykiewicz have developed a model that is global in nature. It focuses on organic pollutants and uses hexachlorocyclohexane as a model organic compound. The model incorporates interactions among the atmosphere, hydrosphere, and lithosphere. In terms of the hydrosphere, it includes ice and snow. In terms of the lithosphere, it includes soil. All three of these are important because they are responsible for sorption of organic components, some of which are released later. The release of previously sorbed components is important to include in modeling because it will complicate modeling and sampling. Verification of this model can be done by sampling. It is a particularly useful model in that it takes into account all the components of the entire environment [16].

7.10. CONCLUSIONS

All types of models—physical, mathematical, chemical, statistical, dynamic, computer, and whole Earth—can be useful in a sampling situation. Such models can predict the fate of components in the environment. They can indicate the most important places and how often sampling should be carried out. Such models can be useful in explaining sampling strategies to both regulatory agencies and interested parties, including the general public. Implementation of a modeling program can be very beneficial to any sampling project.

QUESTIONS

1. Describe two different types of physical models of the environment and tell what characteristics they best model.
2. Discuss the advantages of quantification vs. qualification in sampling.

3. Explain the concept of evapotranspiration and how it can affect sampling.
4. What types of chemical phenomena are used in modeling the chemical activity of components of the environment?
5. Simple mathematical models are often based on statistics. Explain.
6. Explain why computer models are useful in environmental modeling and relate this to sampling.
7. There are two basic landscape models. What are they and how does each account for the movement of soil and pollution?
8. What do **A** and **T** refer to in USLE and how are they related?
9. Explain how particles move through air, water, and soil. How is movement in soil different from that in water and air?
10. Make a simple dynamic model for the fate of nitrate in the environment. (Hint: see Figure 7.5.)
11. Make a complex dynamic model for the fate of nitrate in the environment. (Hint: see Figure 7.5.)
12. Write a simple equation for the future level of potassium in a body of water.
13. Look up the word colloidal as it applies to particle size. What is the upper limit to colloidal-size particles? Why do they not settle out of suspension?
14. Explain how colloidal particles can affect the components dissolved in water.
15. A commercial environmental model can map horizons in soil and layers in the regolith. The program can also estimate the depth to each horizon and layer between sampled points. Explain how such a representation may be used in sampling.

REFERENCES

1. Brady NC, Weil RR. The Nature and Properties of Soils. Upper Saddle River, NJ: Prentice-Hall, 1999:223–234.
2. Svirezhev YM. Thermodynamics and theory of stability. In: Jørgensen SE, ed. Thermodynamics and Ecological Modeling. New York: Lewis, 2000:115–132.
3. Mattigod SV, Zachara JM. Equilibrium modeling in soil chemistry. In: Bartels JM, ed. Methods of Soil Analysis: Part 3—Chemical Methods. Madison, WI: Soil Science Society of America and American Society of Agronomy, 1996:1309–1358.
4. Reichman R, Wallach R, Mahrer Y. A combined soil–atmosphere model for evaluating the fate of surface applied pesticides. 1. Model development and verification. Environ Sci Techn 2000; 34:1313–1320.

5. Leonard RA, Knisel WG, Still DA. GLEAMS: Groundwater loading effects on agricultural management systems. Trans ASAE 1987; 30:1403–1418.
6. Tsiros IX, Ambrose RB. An environmental simulation model for transport and fate of mercury in small rural catchments. Chemosphere 1999; 39:477–492.
7. Stewart MA, Jardine PM, Barnett MO, Mehlhorn TL, Hyder LK, McKay LD. Influence of soil geochemical and physical properties on the sorption and bioaccessibility of chromium III. J Environ Qual 2003; 32:129–137.
8. Universal Soil Loss Equation. http://soils.ecn.purdue.edu/~wepphtml/wepp/wepptut/jhtml/usle.html.
9. Revised Universal Soil Loss Equation. http://www. sedlab.olemiss.edu/rusle/.
10. Brady NC, Weil RR. The Nature and Properties of Soils. Upper Saddle River, NJ: Prentice-Hall, 1999:668–707.
11. Wind Erosion Simulation Models. http://www.weru.ksu.edu/weps.html.
12. Revised Wind Erosion Equation. http://www.csrl.ars.usda.gov/wewc/rweq/readme.htm.
13. Zannetti P. Particle modeling and its application for simulating air pollution phenomena. In: Melli P, Zannetti P, eds. Environmental Modeling. Boston: Computational Mechanics Publications, 1992:211–241.
14. Deaton ML. Winebrake JJ. Dynamic Modeling of Environmental Systems. New York: Springer-Verlag, 2000.
15. Environmental Modeling with GIS. Goodchild MF, Parks BO, Steyaert LT, eds. New York: Oxford University Press, 1993.
16. Koziol AS, Pudykiewicz JA. Global-scale environmental transport of persistent organic pollutants. Chemosphere 2001; 45:1181–1200.

8

Sample Transport and Storage

As soon as the sample is removed from the medium in which it is located it is considered to be in transit. During this process there are three main concerns. First is the integrity of the sample. The sample must be secure from loss, tampering, or contamination. Second, its identification must be maintained so that the results of analysis can be related back to the field and site where the sample was taken. Third, the sample must not be allowed to contaminate or be contaminated by the environments through which it travels. Particular concern is for the personnel handling the sample, some of whom may know little about the potential dangers in the sample.

Next to the actual taking of a sample, its transport and storage are most likely to cause variation or inaccuracies in the analytical results. As mentioned both in previous chapters and below, exposure to light, variations in temperature, wetting, and drying are all storage and transport problems likely to cause variations in analytical results. It is not sufficient to throw the samples in the back of the truck and get around to sending them off for analysis when time permits.

A flow chart of the movement of sample from field to laboratory and analysis is given in Figure 8.1. This is given as an example only, as various analytical laboratories and different projects may have different flows of samples.

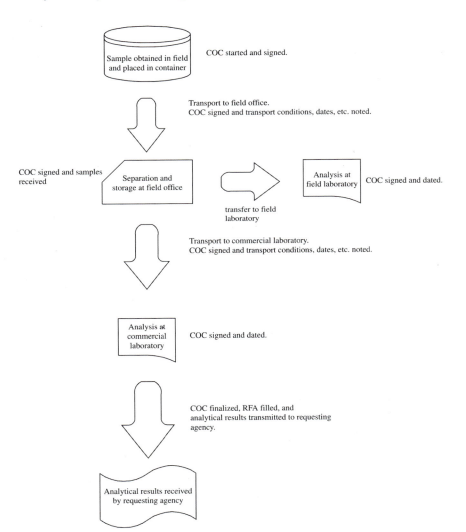

FIGURE 8.1 Flow chart of movement of samples from field to commercial laboratory to results returned to requesting agency. COC is the chain of custody. RFA is the request for analysis.

8.1. SECURITY

There are many different components to maintaining the security of a sample. One is that the sample container must not be broken or tampered with. This means that it cannot be opened at any stage between the initial placement in the sample container and the final opening of the container to take a sample for analysis. If there is a particular concern in this regard, some method of making the sample containers tamper-proof should be found. One approach to this problem is to use fragile labels that are applied to a sample container when filled out, so that tampering will damage them. If the individual containers are not made tamper-proof, boxes or other containers of lots of samples can be made tamper-proof by the use of seals or locks that indicate when they have been opened.

Another area of security is that the sample numbers be secure so that a sample can always be identified. Sample identification numbers must be placed on the containers in such a way that they cannot be erased or removed. Sometimes a little thought beforehand is helpful in avoiding the loss of numbers. Labels inside a sample container with the sample are not recommended in any case because they are a source of contamination and often decompose (e.g., paper labels placed inside containers of moist soil will often degenerate in several days). Under certain conditions labels on the outside can also be lost. For instance, labels placed on the outside of soil drying cups often fall off during drying. Lot containers, however, such as coolers used for shipping samples, will have labels and other paperwork inside them. These are sealed in plastic bags and do not come in contact with the samples.

There are several ways to avoid many labeling problems. Choose labels with high-quality adhesive and write on them with indelible, nonsmearing ink. Place labels on areas of the sample container that will receive little or no abrasion. One interesting place for a label is on the bottom of sample bottles or containers. Often the middle of the bottle bottoms are indented, and thus not subject to abrasion. A second reason this is a good place is that dirty, wet fingers, water, and solvents can get on the top and sides of bottles but often do not reach bottle bottoms. Some sample bags have fold-over tops secured by tabs. The folded area is a good place to put a sample label. A third approach is to use permanent markers to directly mark the sample container.

Samples must be secure from pests. During transport and storage in the field, samples may attract rodents and other animals that can destroy or contaminate samples. Even samples kept for only a short time may be attractive nesting and hiding places for rats and mice. Rodents will attract snakes, and soon there is a whole host of animals in with the samples. Even without opening the sample containers such intrusions may cause contamination because some sample containers are porous and can allow contaminants from animals to enter.

Samples will need to be protected from the deleterious effects of microorganisms in addition to large animals. Because microorganisms can rapidly change the nature of components in a sample, controlling their activity is important. Microbial activity is minimized at low temperatures, and so storage at 4 °C is sometimes required. Also, limiting the time between sampling and analysis will limit the development of this type of problem.

Sample security means that samples are not exchanged for one another and two different batches of samples are not intermixed. Samples must have designated space during transportation and storage. Keeping samples in their designated transportation and storage space will assure that they are not mixed with other samples, which is particularly important when different samples or different sample batches require different analysis or different analytical procedures. Once a sample has started an analytical procedure—extraction, for instance—it is usually not possible to use that same sample or extract in other analytical procedures.

At every step between the field and the final analysis an accounting of where the sample is or has been needs to be kept. It is essential that there be no doubt about the authenticity of the sample. It must be impossible for a sample to be contaminated, switched for another sample, or mixed with another sample. In part this means that samples are never left in the hands of persons not responsible for their transport and storage. It also means having a detailed, well-documented chain of custody (discussed below).

Security can also be thought of in terms of not allowing any sample to exit the container and contaminate or be contaminated by its surroundings. This is dangerous for the persons handling the samples and leads to contamination of the sample, making it useless. Great care must be taken in making sure that the sample is well sealed in the proper container. When several containers are packaged together this means that they cannot interact with each other, either to break open an adjacent container or to exchange components. Often a good approach is to have the samples double-contained. This would be a plastic bag inside a plastic bag or a bottle inside a plastic bag. In addition, the containers packaged together should be cushioned from each other by packing material. A summary of security concerns is given in Table 8.1.

TABLE 8.1   Security Considerations for Field Samples

Container integrity	Handling and package containers to minimize breakage
Sample numbers	Apply sample numbers in several places using high-quality labels and permanent markings.
Pests	Samples must be secure from pests.
Sample mixing	Set up protocols that prevent mixing of samples from different batches or mixing of batches.
Sample leakage	Package samples to prevent leakage during transportation and storage.

8.2. CONTAINERS

Sample containers must have three characteristics. First, the container material must be compatible with the sample and the component or components of interest in the sample. The container must not take away from or absorb any component in the sample, and it must not add any component to the sample. This could be considered negative and positive contamination of a sample. Characteristics of sample containers are covered in more detail in Chapters 3 and 5.

The second characteristic is that the sample container be chosen so that it will not lose components during shipment or transport. Losing components might take place by breaking the container or by diffusion of material out of the container. Diffusion could be through the container walls, which is a common problem with plastic containers. Loss might also be through septa in the tops of bottles. To assure that these types of losses do not take place, both positive and negative control samples can be prepared in the field at the time of sampling. If transportation and storage are of particular concern or are particularly challenging, positive and negative standards or controls can be processed through transportation and storage before actual samples (called transport controls). In this way it is possible to determine if there are critical steps that need to be changed or time constraints that need to be applied to ensure sample validity.

Sometimes samples will come under rigorous conditions during transport and storage. Rigorous conditions are most often changes in pressure and temperature. Samples taken at high altitude or in deep water and analyzed at low altitude or on the water's surface will undergo significant pressure changes. Temperature changes cause pressure changes in the sample container. If the sample contains volatile components or volatile organic compounds (VOCs), component vapor pressures will change with temperature, and appreciable pressure can build up when the sample is obtained under cold conditions and is subsequently warmed to room temperature. Warm samples will lose components when they are opened and the pressure is released. Likewise, a sample obtained at warm temperatures and stored at cold temperatures can have considerable vacuum buildup in the container. Opening the container while it is cold can result in contamination by material moving into the container. Desert temperatures well over 40 °C and arctic temperatures well below 0 °C can be encountered while sampling. Samples taken under either of these extremes will undergo significant pressure changes when brought to room temperature.

The third characteristic of a container is that it must not allow contamination of its surroundings during transportation. This means that the outside of the container is clean and that as discussed above, no part of the sample should be able to exit the container. A common loss would be via breakage, which can be controlled by having sufficient and proper packing material in the shipping container.

8.3. CONTROLS

To make sure that no changes occur during transport, storage controls are used. Common types of controls are listed in Table 8.2. One type would simply be a known; that is, a sample with a known concentration of the component of interest is transported and stored with the test samples. Any change, loss, or gain would indicate a potential problem with the transport

TABLE 8.2   Types of Sampling Controls

Name	Characteristics
Known samples	A sample of known component concentration of interest is included with samples collected in the field. Inclusion must be done in the field.
Negative/blank	A sample known not to contain the component of interest is included with samples collected in the field. Can be either a field or trip blank.
Positive (spiked)	A sample that contains more component of interest than is commonly found in samples.
Surrogate	A sample containing a compound similar to that of interest is included with samples.
Reference	Soil samples taken from similar soil in a different location known not to be contaminated with the component of interest.

and storage procedures. Three other common types of controls are the negative and positive or spiked (component of interest is added to the sample) and a sample containing a surrogate compound.

A negative control would contain all the components found in the sample except the one of interest and can be of two types—field and transport. A field control follows the whole path of samples, while transport controls are used as checks of the transportation portion of the path of samples to the commercial laboratory. These allow for isolation of the source of contamination problems if they occur. Negative controls are important for two reasons. If they contain any of the component of interest serious contamination has occurred, and this is the most serious concern. Any component detected in the negative control that was not present originally also indicates a serious problem, however, and makes the samples suspect. In such cases the analytical results should only be used after verification by using other samples that they are representative of the field conditions.

A positive control contains the component of interest in a known concentration. If a positive control has less of the component than it did in the field, it has lost material. Again this means that all the samples taken, stored, and transported with this sample have probably lost some of this component. One nice aspect of positive controls is that if they all show a similar loss this can be used to estimate the loss from the samples. In this way the original concentration of the component of interest in the samples can be estimated. The sample information thus obtained may be useful; however, such samples and their analytical results must be used with caution due to estimating the concentration of components of interest using a "fudge" factor.

In some cases a surrogate is added to samples in place of the actual component of interest. This is done when the component of interest is particularly toxic or otherwise hard to handle. The surrogate is similar to the component of interest both chemically and physically. Again it is assumed that if the surrogate concentration is unchanged during storage and transportation, no loss of sample has occurred.

As discussed in previous chapters, controls consisting of samples adjacent to the contaminated area need to be taken and sent through the storage and transportation process. Also included would be samples from similar soils that are not from adjacent areas. Both of these types of controls need to go through the same process.

Another type of control to consider is the archived sample. These are samples from each step or stage of the sampling plan that are stored for later reference. They are extremely useful if a question about analytical results or transport conditions arises later in the plan or in the remediation process.

8.4. TRANSPORT IN THE FIELD

Regardless of the sample—air, water, soil, or other—all are most susceptible to contamination in the field. This happens because the sample is placed in the sample container and is kept in the field with other samples while awaiting transport to the field storage area. During this time samples are subject to light and changes in temperature. If samples are taken while the sampler walks the field and samples are put in a backpack, it may be several hours before the samples get to the field office.

Samples will be subject to jolts and bumping as the person or vehicle carrying them traverses uneven ground between sampling sites. While in the field, samples will be exposed to sunlight and variations in temperature. All of these factors will be controlled once the samples are in the storage facility and boxed for shipment to the analytical laboratory.

One good way to avoid some of these problems is to put the samples directly into the lot shipment box that will be used to transport the samples to the analytical laboratory. The box must be covered to keep out light and should be kept at a constant temperature as much as possible. If cooling is called for, boxes will be filled with ice and kept filled at all times. In this way samples are immediately cooled and stabilized in the field. During this time containers must be kept isolated from each other, usually by packing material, so they do not break.

8.5. TRANSPORT BOXES

Transport boxes are used to contain batches of samples and to transport these batches to the commercial analytical laboratory. In some cases such boxes are best obtained from the laboratory doing the analysis. A box supplied by the laboratory will come with instructions for both packing and the conditions under which the samples must be kept. A typical transport box is shown in Figure 8.2. The box shown contains a number of different containers for different types of samples. Normally a box would only contain the types of containers designated for the type of sample being collected.

Transport boxes are constructed of many different types of materials. Simple cardboard boxes can be used but are not strong, and sample weights must be carefully monitored to make sure that the box does not contain too many samples. Simple wooden, plastic, and metal boxes are also used. These may be constructed with internal sample racks that keep samples protected from each other. Insulated boxes are also used for some samples.

The box shown in Figure 8.2 is a relatively simple insulated plastic cooler. Typically these are maintained at some lower temperature (4 °C is

FIGURE 8.2 A typical sample transport box and various sample containers. The container marked with A is used for checking sample box temperature.

common) from placement of sample in them to removal of sample at the commercial analytical laboratory. The internal temperature will be monitored by having a temperature monitoring container, such as that shown in Figure 8.2 (i.e., the bottle labeled A, in the box).* Measuring the temperature of the bottle contents upon arrival at the laboratory serves as a check on the temperature of the sample box contents. If the temperature is higher than required when the sample reaches the commercial laboratory, the sample may be rejected or is at least suspect. It is also possible to have electrically refrigerated sample boxes that have small electrical or mechanical cooling mechanisms and that can plug into an electrical outlet or a vehicle electrical system. This type of sample transport container will maintain the correct temperature in hot conditions as long as there is a source of electricity.

---

* In many instances temperature monitoring will be required by regulatory agencies.

Various paper forms will come with the shipping container and must be filled out and enclosed with the samples when they are returned to the analytical laboratory. Typical forms are shown and discussed in Chapter 10.

8.6. TRANSPORT OUT OF FIELD

Care must be taken in transporting samples out of the field to the field storage facility. Samples should not be thrown in the back of a pickup and left for several days before being placed in the field office storage facility. This is of particular importance, because samples banging around in the back of a pickup are prone to both breakage and picking up contamination, as well as being exposed to light and variations in temperature. Any of these variations alone or in combination can cause changes in the component or components of interest.

8.7. STORAGE

There will almost always be time between taking a field sample and its field laboratory and final analytical laboratory analysis. This will require a secure sample storage area at or near the field office and at the analytical laboratory. The preference is to have this be a separate building or at least an area separated from most or all of the other activities occurring at the field office. As discussed above, this area must be secure from intrusion by animals and unauthorized intrusion by persons working in the field or in the field office. It must also be free of possible contaminants. Using a room that has previously been used for the storage of feed, fertilizer, insecticides, solvents, or some unknown purpose is not acceptable. Equally important, the stored samples must not be located in a way that allows materials from samples to contaminate either the office area or the personnel working there.

The room should be both humidity- and temperature-controlled. In some instances this will require only heating or cooling, while in others maintaining a constant temperature may require both a heating and a cooling unit. Constant temperatures are often best achieved by having a heater and a refrigeration unit working against each other. In high humidity areas removing some humidity may also be advisable. Air conditioners will condense moisture out of the air, but this water must be drained to the outside of the storage room.

The storage room may need to have several sections or different areas, depending on the storage requirements of the samples taken and the components of interest they contain. Some samples may need to be stored at 4 °C, while others may need to be kept at −20 °C. Yet others may need to be kept at room temperature. Also, some samples may need to be stored dry, while others need to have the original moisture content maintained. Ross and Bartlett [1] found that nitrate levels increased dramatically in forested Spodosols stored at 3 °C after only 24 hr. On the other hand, Martens and Suarez found little change in selenium species in air-dried soil samples. They found significant changes in selenium oxidation states of air-dried alfalfa (*Medicago sativa*), however [2].

Another aspect of storage temperatures is the effect on the physical condition of the sample. Compounds in solution may precipitate at reduced temperatures and not go back into solution when warmed. With some hydrocarbon samples cooling may result in separation of various phases. This separation may not be reversed by heating, and may produce analytical results that do not provide the information needed [3].

Some monitoring of the sample room needs to be done. The minimum acceptable monitoring is to record the temperature and humidity in the room every time a sample or batch of samples enters or leaves. This should be done on a sheet kept in the room. Each different area of the room needs its own sheet, and all sheets must be filled in whenever a sample is placed in the room, no matter which section it is placed in. Recording thermometers and humidity indicators can be installed in such a way that a continuous record of the room's conditions can be kept. If this approach is taken, a separate instrument is needed for each area of the storage room. Alternatively, sensors could be connected to a computer and the computer could be used to record information about the room. A connection can be made that is wireless or with wires. A single computer can receive and store signals from many different sensors in different parts of the storage room. Also, the computer can be set up to notify personnel when the storage room or any part of it is not at the proper temperature or humidity. The storage room or area thus needs to be versatile enough to allow for all the different storage requirements of different samples and components of interest.

Locks on the doors will provide security, as will motion sensors and cameras so that all movement into and out of the room is controlled and recorded. This provides added security for the samples.

8.8. ARCHIVAL SAMPLES

It may be desirable to save some samples, called archival samples, until the end of the project or even longer. They must be kept separately from all other samples and labeled clearly as archival samples. These are duplicates or part of samples taken at a particular time during sampling of a field. They can be used to check the accuracy and authenticity or estimate the original condition of the samples at a later time during the project or even years after the termination of the project. It must be kept in mind, however, that the samples and the analyte of interest may have changed significantly during storage. For this reason any use of archival samples in comparison studies must be done with extreme caution.

8.9. SEPARATION OF SAMPLES AT THE FIELD OFFICE

Field analysis will be done in the field by the sampler or sampling crew. In some cases this may not be feasible and the samples will be taken to the field office laboratory for analysis, and some or all of the sample or samples will be analyzed and the rest shipped to the commercial analytical laboratory. Separation is best done as the samples enter the storage facility by immediately placing those to be analyzed in a different place from the samples to be shipped to the analytical laboratory. Samples will be analyzed immediately, and thus will not need storage.

8.10. FIELD LABORATORY HANDLING

The analysis to be done immediately in the field or at the field laboratory is given in Chapter 10. This includes such things as pH, electrical conductivity (EC), and dissolved oxygen in the case of water samples. For pH and EC the analysis can be done directly on the samples in the field as they are taken; that is, the samples need not be dried and sieved. In some cases, however, it may be desirable to bring samples to the field laboratory to sieve if they contain a lot of plant material or large rocks, gravel, and so forth; otherwise a minimum of handling is recommended.

Although the above analyses are straightforward and not delicate, care still must be taken to obtain accurate results. The word delicate is used to indicate that small changes in the way the analysis is carried out will usually not result in a change in the results obtained. For instance, slight dilution of the highly buffered solutions used for standardizing a pH meter will cause little, no, or insignificant changes in the measured pH. This does not mean that care need not be taken in doing the analysis, however. It does.

8.11. TRANSPORT TO THE COMMERCIAL LABORATORY

A chain of custody (COC) form and other forms (see below) as required by the laboratory must be started in the field as the samples are taken and be continued by the receiving laboratory. Before shipping, however, the analytical laboratory is notified because it needs to be ready to handle and store the samples in such a way that the components of interest are preserved.

Commercial carriers are usually used to transport samples to the laboratory for analysis. An ideal situation would be one in which the commercial analytical laboratory is close enough to the field being sampled that the samples can be locally transported. This situation does not often happen, and most often the samples will need to be shipped to the laboratory using a commercial carrier. The carrier should be chosen so that the transit time is no more than 1 day, if at all possible. Usually the project manager at the analytical laboratory will be able to suggest carriers and shipping requirements.

Just as individual samples must have traceable numbers, so also must lots have traceable shipping numbers. Registered or certified mail is traceable, but may not be sufficient for the numbers and weights of samples that need to be shipped. Companies such as Federal Express or UPS assign traceable numbers to shipments and will often pick up samples at a designated location. This decreases handling and thus increases security. Using one of these carriers the progress of the sample from pickup to delivery is traceable, accessible, and can be followed using a computer and the Internet.

In many cases shipment will involve air transport. Very specific restrictions are applied to samples shipped by air. Different types of samples (i.e., samples containing different types of contaminants) will have different restrictions. For instance, there are specific requirements for packaging oil-contaminated samples. One of these is that the inside of the box be fitted with an oil-resistant liner. There are also limits on the size and weight of sample containers. All these restrictions must be strictly adhered to when shipping by air.

The shipment must be accompanied by a COC form updated by various persons handling the shipment along the way and signed by the appropriate persons in the appropriate places. This is particularly true if a national carrier such as UPS is not used to ship the sample. This may seem simple, but a truck driver who takes the samples to the local airport will handle samples, and this person needs to be known. Likewise, the shipping department at the air shipment office will receive the samples and perhaps hand them off to another person to get them to the airplane. Indeed, there may be three or four different people handling the box between its arrival and being placed on the airplane. When the plane lands, the same or a greater number of people may handle the sample box before it gets to the laboratory receiving office. All of the people handling the container of samples need to be identified, and the length of time the sample is in their possession must be known. The greater the number of people handling the sample box the greater the chance that something will happen that will render the samples unusable.

8.12. STORAGE AT THE COMMERCIAL ANALYTICAL LABORATORY

Samples will be stored at the analytical laboratory, sometimes for very short periods of time. Normally the laboratory will have standard operating procedures for sample handling and appropriate storage facilities. A visit to the laboratory to see how samples are handled is recommended. It is always advisable to ascertain how long and under what conditions samples will be stored at the laboratory between receiving and analysis. Second, a quick walk-through of the facility is necessary. Samples lying around in a haphazard fashion in open containers are cause for concern. Such circumstances lead to contamination and cross-contamination of samples.

A third concern is maintaining the proper environmental conditions during this time. If the sample temperature is to be maintained at 4 °C, the appropriate cold room should be in evidence. In some cases samples may need special environmental conditions for storage, and the laboratory must be willing and able to provide these conditions.

8.13. CHAIN OF CUSTODY

The chain of custody (COC) is the most important component of the secure movement of samples from the field to the laboratory to final receipt of the analytical results. Another component associated with COC is the letter of transmittal. Such a letter is not typically a COC, in that samples may have both a COC and a letter of transmittal. Also, a letter of transmittal can mean different things to different people; a request for a new position in an organization may be called a letter of transmittal, for example, and thus many of these types of letters are unrelated to samples or safe sample movement. Although COCs may be thought of as being associated with evidentiary forensics samples, they are equally valuable and important in environmental and geological sampling.

The COC accounts for where the sample originated, where it has been, who has handled it, and what conditions it has been exposed to. It should prevent loss, misplacing, and tampering with the sample in any way. It will also help prevent samples from either being analyzed for the wrong constituents or being subject to the wrong procedures. To accomplish this it is best to minimize the number of people involved in the COC and sample handling. The more people handling a sample or container of samples, the greater the likelihood of contamination, loss, or mislabeling of the samples. If one person is sampling and is also responsible for storage, that person does not necessarily need to sign the COC each time he or she handles the sample. Each time the sample or samples are transferred to a new location, however, this needs to be fully noted. The dates, times, and conditions of transfer need to be recorded, even if the same person is doing all the work [3].

The COC starts in the field with taking the sample and putting it in the sample container. The sample is recorded in the project notebook, given a number, and recorded on a COC form. This form then stays with the sample until the analytical results are reported back to the entity requesting the analysis. If bar codes are used, a code label can be placed on the COC form [4]. If a group of samples are to be shipped as a batch, a batch COC needs to be prepared. Individual sample COCs go in the batch box with the samples, and the batch COC is signed as needed by persons handling the box. Again, if bar codes are used a bar code for the box can be prepared and used to keep track of the box. An illustration of the type of information needed for a COC form is given in Figure 8.3. (A more detailed COC is given in Chapter 10.)

At each step of this process a responsible person should be designated to receive samples, sign the COC forms, and put samples in the appropriate storage area or transmittal container. This person will also be the contact person during the time the samples are in his or her possession. In small firms, in which there are a limited number of employees, this may seem unnecessary. It is not; but it need not be a complicated procedure. In large firms it is essential that persons handling samples be identified and all the time the samples that are in their possession be accounted for.

8.14. COMPUTER CHAIN OF CUSTODY

A COC can and should be set up using computers. This should be done even though a detailed paper COC is also maintained. One advantage of a computer COC is that it can be quickly accessed from many locations when a problem occurs. Software can also allow for the insertion of additional written descriptions and explanations in such a way that they do not interrupt the flow of the COC because they are readily available as "pop-up notes" that appear with the push of a button. This is much more useful than shuffling through numerous pages of information to find a needed explanation or description. The information in the pop-up could be about storage facilities, changes in temperature, or other conditions that might

Field Sample
Chain of Custody Form

Page_____of_____pages

Person originating form _____

Origination date/time _____   Number of samples _____

Originating organization _____   Sample media _____

Sample numbers inclusive

_____

Ultimate receiving organization _____

Job number _____

Person obtaining samples—name, date, and time	Receiving location	Deposition location	Person accepting delivered samples—name, date, and time

FIGURE 8.3 A suggested chain of custody form for field samples.

affect a sample's integrity, and needs to be included with the COC. Software specifically designed for COCs is available [5,6].

An additional advantage is that computers can be accessed from remote locations if they are connected to the Internet. They can also distribute or receive information from several locations simultaneously. This means that the original COC can be accessed from either the field in which the sampling is occurring or from the analytical laboratory in which the analysis is occurring. It also means that at any time during transport changes in the COC can be transmitted to both the field and to the analytical laboratory.

With computers there is always the danger of data loss. Acquiring and using an uninterruptible power supply (UPS) is a good way to help prevent data loss. These power supplies will keep the computer running for a short period of time, which is usually long enough to allow data to be saved, and the computer can be turned off when there is a power failure. A safer strategy, however, is to constantly and conscientiously back up data. Keep in mind that a UPS unit does not preclude the need for daily backup; it can fail and it cannot prevent failure of other components of a computer.

There are five major ways to store data; 3.5 floppy disk, zip drives, hard drives, CDs, and DVDs. A 3.5 floppy disk can be used and has the advantage that it can easily be carried from place to place and can be kept with the COC. These disks, however, are limited in the amount of data they can store. A hard drive is a convenient storage device and can typically hold large amounts of date. Hard drives are usually internal to the computer, however, so either data are transferred to the hard drive when the office receives then or the computer must go with the sample. This obviously will not work with a desktop computer containing a hard drive, but is feasible with a laptop. There are external hard drives accessible through a computer's USB port. This hard drive can be carried from computer to computer, but transport is not easy because of its nature and its susceptibility to damage and data loss.

Either internal or external zip drives are also available. These drives and their removable disks hold a moderately large amount of data, and the disk can easily be transported or kept with the COC. External zip drives are small enough to carry around conveniently, and if they connect via a USB connector, can be connected to available computers at any location. One important drawback to all the storage devices discussed so far is that the data on them can be lost or erased.

For a permanent record, a CD or DVD recording can be made of the COC and the accompanying information (usually referred to as burning). For a permanent record a CD-R or DVD-R, ("read only") must be used. There are CDs that can be rewritten, but for COCs these are not recommended. Once a CD-R is burned, it is a permanent record of the COC and accompanying explanations. A CD player/burner or DVD player can be bought as an internal or external unit. An external unit would then be available to carry between computers. DVDs hold a great deal more information than CDs, so a single DVD may be all that is necessary to store all the information gathered about a particular field, including maps and pictures.

Newer types of storage devices, flash memory, memory sticks, and flash cards, are also available. These are used extensively on digital cameras, can also store data and are relatively small. They hold about the same amount of data as a zip disk; that is, up to 256 megabites (MB) in a very small unit. They are particularly useful if a number of pictures are to accompany the sampling plan, and can easily be carried from place to place and used on any number of computers, both desk and laptop. A drawback is that some of these devices require a special reader or port to transfer data from these devices to a computer. Two other potential drawbacks are that some seem to use significant amounts of energy, and this may be a problem when they are used with digital cameras, which have dry cell batteries. The second drawback is that they can be erased, and so data can accidentally be lost. A number of common computer storage devices are shown in Figure 8.4. Table 8.3 lists various storage devices and gives some information about their characteristics.

8.15. REQUEST FOR ANALYSIS

It is essential that samples leaving the field office for the analytical laboratory be accompanied by a request for analysis (RFA) form, which will be different for different laboratories. Both the COC and the RFA forms should have a contact person clearly and easily identified on them. The

FIGURE 8.4 Some examples of computer storage media. From left to right: CD, zip disk, and 3.5 floppy disk.

TABLE 8.3 Computer Storage Devices

Device	Type of media	Condition of stored data
Floppy drive	Floppy disk	Data easily stored and transported. Limited storage capacity. Data can be erased or otherwise lost.
Zip drive	Zip disk	Data are easily stored and transported. Will hold more data than floppy disk. Data can be erased or otherwise lost.
Memory stick and card	Memory chip	Data easily stored and transported. Holds about same amount of data as zip drive, but is much smaller and more transportable.
Hard drive	Nonremovable hard disk	Holds large amount of data. Usually an integral part of a computer. Not easy to transport. Data can be erased or otherwise lost.
CD burner	CD-R	Holds large amount of data. Burner is usually integral part of computer. Data cannot be erased.
CD burner	CD-RW	Same as above, but CD can be rewritten on, thus loss of data is possible.
DVD burner	DVD-R, DVD-RW	Similar to CD, but holds gigabytes (GB) of data. Can either be write once or rewritable.

contact person needs to be identified as to position, mailing address, phone numbers, and e-mail address. E-mail addresses are handy because an exchange of e-mails will often facilitate phone contacts when needed. The reason this is so important is that the laboratory needs to know of whom they should ask questions if there are problems with the samples.

Discussion with the analytical laboratory personnel, particularly the project manager assigned to the project, of the COC and RFA needs of the submitter is advisable and will eliminate confusion in sample transfer, recording COC data, and processing the RFA. It will also facilitate the return of analytical results to the contracting agency. Examples of RFA and results of analysis forms are given in Figures 8.5 and 8.6. These figures do not correspond to any commercial or government laboratory form; such forms will be more detailed. (Additional forms of this type are given in Chapter 10.)

Request for analysis form (not related to any commercial or governmental organization form)			
Analytical laboratory information—name, address, contact person, address, phone numbers, E-mail, etc.			
Submitter information—name of organization, address, contact person, phone numbers, E-mail, etc.			
Requested organic analysis		Requested inorganic analysis	
Analysis requested	Check all that are desired	Analysis requested	Check all that are desired
TPH		Cr	
VOC		Ni	
DAPAL		Cu	
PCB		AS	
Total C		Se	
CHN		Hg	

FIGURE 8.5 Example of the type of information needed for a request for analysis. Actual forms will ask for more information and detail.

8.16. CONCLUSIONS

Handling samples between the field and final analysis is the second potential area of sample loss or contamination. For this reason sample security is essential. Proper containers and controls must be used in sampling and shipping. The conditions that samples are exposed to during shipping and storage must be controlled if the component of interest is to be maintained in its original concentration and form. Chain of custody forms must be scrupulously kept during the whole sampling and transportation process. It is highly recommended that a computer-based COC be maintained, along with a paper COC to accompany the samples. A request for analysis form must accompany the samples to the analytical laboratory. Discussions with the analytical laboratory's project manager assigned to the project is essential for successful sampling and analysis of samples.

Request for analysis results form (not related to any commercial or government form)									
Analytical laboratory information—name, contact person, address, phone numbers, E-mail, etc.									
Submitter information—name of organization, contact person, address, phone numbers, E-mail, etc.									
Field sample number	Lab. sample number	Sample amount	Sample condition	Component analysis requested					
				Cr	Ni	Cu	As	Se	Hg
AC13a									
AC13b									
AC13c									

FIGURE 8.6   Example of information provided in a request for analysis results form. Actual forms will ask for more information and detail.

QUESTIONS

1. Explain three reasons why security of samples is important. What security precautions are needed?
2. Describe the characteristics that a sample container must have. What must the container do and what must it not do?
3. In terms of transporting samples, what considerations about sample containers should be taken into account?
4. What are two types of common controls and what information about samples and sample handling does each provide?
5. Explain how samples can be contaminated in the field.
6. What characteristics must a sample transport box have?

7. Describe the characteristics that a sample storage facility must have.
8. What steps are followed in separating and handling samples at the field sample laboratory?
9. What are archival samples and why may they be important? What factor must be remembered if the archival samples are used to derive data to compare with the current samples?
10. Describe the characteristics of a chain of custody.
11. What information is called for in a request for analysis?
12. What is the difference between a chain of custody and a letter of transmittal?
13. What are the advantages of using computers in maintaining a chain of custody?
14. Using the Internet, look up the shipping requirement for air shipment of water or soil contaminated with oil or hydrocarbons.

REFERENCES

1. Ross DS, Bartlett RJ. Effects of extraction methods and sample storage on properties of solutions obtained from forested spodosols. J Environ Qual 1990; 19:108–113.
2. Martens DA, Suarez DL. Changes in the distribution of selenium oxidation states with sample storage. J Environ Qual 1997; 26:1711–1714.
3. Marine Safety Laboratory. http://www.rdc.uscg.gov/msl/downloads1.html. Pp. 8–13.
4. Bar Code Primer—Introduction to Barcoding. http://barcodehq.com/primer.html.
5. Murphy C, Briggs P, Adrian B, Wilson S, Hageman P, Theodorakos P. Chain of Custody—Recommendations for Acceptance and Analysis of Evidentiary Geochemical Samples. U.S. Geological Survey circular 1138. Washington, DC: United States Government Printing Office, 1997.
6. e-Chain. http://www/enabl.com/p_echain.html.

9

What Is Present

Analytical results having unexpected amounts or types of components are not necessarily in error

The environment is made up of the atmosphere, hydrosphere, and lithosphere. The atmosphere is relatively simple and its composition is relatively stable. Only those elements, compounds, ions, and molecules that are gases or that can remain suspended in the atmosphere will stay there for a significant period of time. The hydrosphere is more complex, having various levels of salts and other compounds that originate in the lithosphere containing it. The lithosphere is far more complex and can contain all the elements, compounds, molecules, and ions present in the Earth. The three parts of the environment cannot be separated from each other. Components occurring in one sphere can and do move into the other spheres.

The three spheres are sampled for contaminants on a regular basis. For the most part components occurring in the atmosphere are either its natural components or components put there as a result of the activities of man. With few exceptions, such as reactions among chlorine, ozone, and nitrogen oxides, the number and complexity of reactions producing compounds, ions, and molecules in the atmosphere are both limited and relatively simple.

Both the hydrosphere and lithosphere, particularly the soil, are different. Soil is intimately connected to water because rain falls on it, infiltrates, percolates, and finally is released into bodies of water. During these processes water picks up molecules, ions, and elements from the soil. The soil is a great reactor, producing all manner of inorganic, organic, and biochemical molecules, as well as ions, some of which will be dissolved by the water moving through it. The question then becomes what types of compounds are common and commonly produced in this reactor.

Added to the natural components produced by normal soil reactions will be components produced by man and either unintentionally or intentionally added to soil and water. Both these groups of compounds are important because they will show up in samples taken from soil and water and because it is important not to confuse naturally made and occurring components with contaminants added to soil and water by man.

Soil is a mixture of inorganic and organic solids plus a solution of inorganic and organic ions, molecules, gases, and elements in water. It also has a gas phase made up of nitrogen, oxygen, carbon dioxide, water vapor, argon, methane, and a number of other minor gaseous constituents. Many think of the inorganic fraction as being inert and unchanging. This is not true. The variety of inorganic components, their structure, and their composition is vast. Likewise, the organic fraction is large and immensely varied. This in turn leads to a complex solution of these components in water and the occurrence of many unexpected components in the environment.

The inorganic phase is mostly silicon, aluminum, and oxygen. The organic fraction is made up mostly of carbon, hydrogen, nitrogen, and oxygen. Neither of these characterizations provides any useful information about the composition and activity of soil, however. Even small amounts of clay particles, particularly the very active clay fractions, can make a large difference in a soil's characteristics. This includes how easily a soil is sampled. A sandy soil with small amounts of clay can be sampled using a core sampler. Sandy soils without any clay must be sampled using a bucket-type sampler. The same thing can be said for the other components in soil, including organic matter.

The soil solution is made up of water. In most cases it contains a complex mixture of inorganic and organic ions and molecules and dissolved gases. Relatively small changes in any of these constituents can have a pronounced effect on the soil solution's characteristics. At pH 7 the soil solution contains 0.0000001 moles of H^+ per liter.* This is not very much! A

---

*It is more accurate to call this the hydrogen ion activity in soil rather than concentration; however, for our purposes here concentration illustrates the point better.

slight increase in H^+ to 0.000001 moles/liter makes the solution acidic, while a change to 0.00000001 H^+ moles/liter makes the soil solution basic.

In like fashion, small changes in the gaseous content of soil can dramatically affect its characteristics. An increase in the carbon dioxide concentration in soil air will change the soil solution pH. It can also cause the precipitation of various ions as carbonates.

All of this leads to three important considerations in sampling. First, changing soil conditions will change the amount of a component that is found. In some cases it may even determine if a component is found or not. Because such a variety of components is present, sampling must always be accompanied by controls. It is not good enough to simply say that lead, chromium, arsenic, or any one of innumerable other inorganic and organic compounds is found in a soil sample. The concentration at which it is found is critical because it must also be determined if this level is natural for the soil or not. Is the component significantly above background levels or not? Another important question concerns the biological availability of the component. Is the component in a form that is biologically available or unavailable? Only when these questions have been answered can one begin to assess the need for further sampling and, potentially, remediation.

9.1. IN TOTAL

The three spheres of the environment have very different compositions when looked at from the perspective of gas, liquid, and solid phases. The atmosphere is predominantly a gaseous phase. Relatively small amounts of liquids and solids are suspended in it. The hydrosphere is predominantly a liquid phase. It has variable amounts of salt, relatively small amounts of gas dissolved in it, and small amounts of solid suspended in it. On the other hand, soil has relatively large amounts of gaseous, liquid, and solid phases in intimate contact with each other.

A sample of soil has three phases—solid, liquid, and gas—intimately mixed together. It is roughly half solid and half void space. The solid phase in many soils is 98% inorganic and 2% organic, or 49% and 1% of the total volume. The void space is roughly half air and half filled with water. (See Figure 9.1.) The inorganic component has an elemental composition of aluminum, silicon, iron, titanium, and oxygen if only the major components are considered. The relative amounts of these three and other major elements change as soil develops.

As can be seen in Figure 9.2, rock is composed primarily of silicon and aluminum. It also contains appreciable amounts of iron, potassium, and sodium with smaller amounts of calcium and magnesium. The more soluble

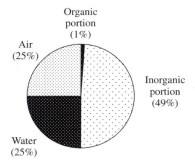

FIGURE 9.1 Percentage composition of a soil sample by volume. Water and air contents vary with moisture content.

components are leached out of the soil parent material or regolith over time and the more resistant materials are left behind. In the pie chart of weathered rock there is thus an increase in aluminum and a decrease in silicon content. Iron and magnesium increase, in concentration, and calcium, potassium, and sodium decrease. Along with this is a dramatic increase in the amount of water held by the inorganic components. The composition of the inorganic phase of soil is thus constantly changing, and the relative amounts of components present will be different at different times.

The organic constituents are composed of carbon and hydrogen with some nitrogen and oxygen. On a weight bases the hydrogen is not appreciable; its atomic mass is 1 atomic mass unit (AMU). On a molar basis, however, it is extremely important because there are approximately 2 moles of hydrogen for every mole of carbon.

The soil solution phase is primarily water or H_2O. In this case elemental analysis would show this phase to be made up primarily of oxygen with some hydrogen. All other constituents would be at low concentrations.

The gaseous phase of soil is mostly nitrogen (N_2), oxygen (O_2), and water. Because of its low concentration, carbon from carbon dioxide or other sources might not show up as being important. Methane and hydrogen may also be present in low concentration in flooded soils. This type of description of the soil atmosphere is similar to that commonly given for atmospheric air, which is said to be composed of nitrogen, oxygen, and carbon dioxide. Nitrogen represents approximately 80% of air and oxygen 20%, with carbon dioxide being 0.003%. Note that argon, which is never mentioned, makes up 1% [1]! Additionally, water can be as much as 5% of the soil atmosphere.

Cationic composition of unweathered rock
(excluding the O anion which is most common)

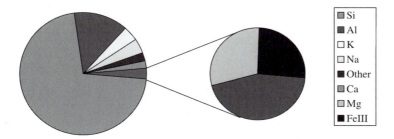

Inorganic composition of weathered rock

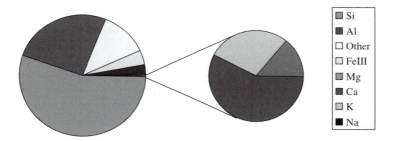

FIGURE 9.2 Inorganic composition of rock and soil before and after weathering. Data taken from Ref. 10.

9.2. THE INORGANIC COMPONENTS

When discussing the specific inorganic contents it might be good to start with those present in elemental form. The gaseous phase contains nitrogen, oxygen, argon, and in the regolith, radon. Both nitrogen and argon are considered to be inert in the environment. On the other hand, oxygen is very reactive and necessary for respiration by most organisms and plant roots. Helium and radon can also be found in the soil atmosphere and are particularly apparent near geologic fault lines [2,3]. Even mercury gas (Hg gas) has been reported in the regolith atmosphere [4]! There is little in the way of elements in the hydrosphere, soil solution, and water in either the regolith or in rock except for the dissolved gases nitrogen, oxygen, argon, hydrogen, and so on. In some cases colloidal-size carbon particles may be present and may be of particular concern because of their high sorptive capacity [5].

The solid phase of soil, regolith, and rock will contain very small amounts of noble metals in the elemental state [6]. There will be some elemental carbon present, much of it as a result of fire, either natural or man-made [5]. There may also be some gaseous elements trapped in solid matrices in soil. The overwhelming amounts of the inorganic components are in the form of ions and molecules.

In some places the soil, regolith, and rock may contain arsenic and uranium. Radon when found comes from the decomposition of uranium in the regolith. It is of note because of its diffusion into house basements [7–10].

9.2.1. Elements in Combination

Many more elements than those mentioned above are commonly found in the environment but are present as compounds or ions rather than in the elemental state.

Of the 110 or so elements in the periodic table, 36—or 33%—are common and important in the lithosphere. Their importance comes either because they are essential for plants or are needed in the diets of animals, or because they or one of their forms is toxic. In addition there are other elements, such as silver, gold, argon, and helium, found in low concentrations, that are not essential to plants or animals and are not toxic. Because of their environmental and economic importance all have had specific analytical procedures developed for their detection and measurement.

In Table 9.1 the elements commonly analyzed for are arranged in three columns. The first column shows elements essential to plants and animals. The second shows elements toxic to plants and animals at high concentrations but essential at low concentrations. The third contains elements generally considered to be toxic. Not all elements listed in column 1 are essential for all plants and animals. For instance, cobalt (Co) is generally not needed by plants, but is essential in the nutrition of animals. Likewise, silicon is not essential to animals, but is for some plants. The same thing can be said in column 2. As shown in column 3, some elements, such as aluminum (Al), are toxic to plants but not to animals. On the other hand, lead (Pb) is toxic to humans but is not especially toxic to plants [11]. The last

TABLE 9.1 Elements Found in Soil

Plant and animal essential elements found as ions	Essential and toxic	Potentially toxic	Found in elemental form
Macro[a]	Se	Al	H_2
H	B	Li	O_2
C	As	Ba	N_2
O	Na	Rb	Rn
N		Cs	He
K		Be	Ar
S		Sr	C
Mg		Cd	Cu
Ca		Pb	Pb
Si		Hg	Ag
Fe		Br	Au
Micro[a]		F	Fe[b]
Cr		Rn	
Mn		U	
Cu			
Zn			
Mo			
Co			
Ni			
P			
Cl			

[a]Macro elements are present at relatively large levels and micro are relatively low levels.
[b]Fe only counts if you consider meteorites.

column gives the elements sometimes found in elemental form in the environment.

All the natural elements are found someplace on the Earth. The elements mentioned above, however, are common in all areas, while other elements are found in detectable amounts only in restricted locations.

As stated above, these elements are found as ions combined with other ions in the environment. In the case of inorganic compounds these are generally salts. The metals are found as positive ions, or cations, and the nonmetals as negative ions, or anions, and in some cases both can form negatively charged oxyanions when combined with oxygen. (See below.)

Carbon in organic compounds is not ionic, but forms compounds that do not generally have full positive or negative charges.*

9.2.2. Nonmetal Compounds in Soil

The nonmetals in soil include compounds of carbon, oxygen, nitrogen, chlorine, silicon, phosphorous, sulfur, bromine, iodine, arsenic, fluorine, and selenium. A number of inorganic carbon compounds, such as carbon dioxide and carbonates, occur in soil. The vast majority of carbon compounds in soil are organic, however, and will be discussed in the section on organic compounds below. Oxygen occurs in carbon dioxide, but also more importantly in water. Water is a unique and critical component in the environment and will be discussed in the next section. Oxygen occurs in carbon monoxide and ozone, which is important in both the upper and lower atmosphere. Oxygen also occurs in all the oxyanions.

Nitrogen can occur in several different types of inorganic compounds in the environment. These will commonly be compounds of ammonia, ammonium, nitrite or nitrate, and various nitrogen oxides. Although other nitrogen compounds have been reported in the environment, they are generally in low concentration and do not gain much attention from environmentalists. Any time an unusual amount of any type of nitrogen compound is found in any environmental sample there is cause for concern.

Phosphorus is found bonded to oxygen, forming phosphates. Most often phosphates are formed as combinations with calcium to produce various calcium phosphates. These may also have other elements associated with them, such as hydrogen, fluorine, aluminum, and iron, forming other naturally occurring phosphates.

Both chlorine and bromine are commonly found in the environment. Their concentrations increase as one moves toward the oceans or salt deposits. Both are commonly combined with alkali metals to form salts. These salts are highly water soluble and are easily and rapidly leached from soil during its formation. Eventually these salts are deposited in the oceans [12].

Silicon is almost exclusively found as silicon dioxide (SiO_2) and silicates in the environment. In silicates it takes a tetrahedral form in the center of four oxygens, which bond to other constituents, such as aluminum. In this way it is an integral part of soil minerals in the sand, silt, and clay fractions.

---

* Exceptions to this are organic acids, which can ionize to form anions, and amines, which act as cations when they associate with protons.

9.2.3. Metal Cations

The alkali and alkaline earth elements form simple positive cations. The alkali metals lithium (Li^+), sodium (Na^+), potassium (K^+), rubidium (Rb^+), and cesium (Cs^+) form ions with a single positive charge or oxidation state. The alkaline earth metals magnesium (Mg^{2+}), calcium (Ca^{2+}), strontium (Sr^{2+}), and barium (Ba^{2+}) occur as simple cations with two positive charges or an oxidation state of 2^+. Aluminum is found only in the 3^+ oxidation state. Other metals, such as chromium, manganese, iron, cobalt, nickel, and copper can occur in two or more oxidation states. For these elements not only is their presence important but also their oxidation state. Iron II (Fe^{2+}) is more soluble than iron III (Fe^{3+}). Chromium IV (Cr^{6+}) is considered to be more toxic than chromium III (Cr^{3+}), thus the form or species of an element is as important or more important than simply its presence.

High concentrations of lithium, sodium, and potassium are associated with high soil pH. Normally soil pHs do not go above 9, thus higher pHs may indicate that there are high levels of the alkali metals and that something unusual is happening and needs to be investigated further.

Aluminum is associated with acid soil conditions. Soils with low pHs will be found to have high levels of available Al^{3+}. This is of concern because this cation is toxic to most plants. The exceptions are the so-called acid-loving plants, such as rhododendron and blueberries [13].

Elements found in multiple oxidation states will generally be found in the state appropriate for the conditions existing at that time. Under reducing conditions iron will be found as Fe^{2+}, and under oxidizing conditions it will be in the Fe^{3+} oxidation state. This is generally true for all the cations, which exist in several oxidation states (i.e., lower oxidation state under reducing conditions and higher oxidation state under oxidizing conditions). Manganese is found as 2^+, 3^+, and 4^+ in soil and acts as if it is between oxidation states because of this mixture [14].

Another exception is molybdenum (Mo). It is an exception not because of multiple oxidation states, but because it is found as an oxyanion, molybdate ($MoO_4{}^{2-}$), rather than a cation in soil [15].

When sampling and analyzing soil it is important to keep the oxidation states of the metals of interest in mind. Samples from reducing conditions will typically show higher levels of soluble transition metal cations than those taken from oxidizing conditions. A second component of this is that during the sampling, samples are usually exposed to oxygen. Such exposure can lead to changes in the oxidation states of the metals the soils contain. This would also happen if these samples were exposed to air during storage or transport. Oxidation at any step can have a pronounced effect on the analysis and the apparent biological availability of a metal.

9.2.4. Nonmetal Cations

There are two common nonmetal cations found in soil. The first is the H^+, a proton, which does not really occur as H^+ but as the hydronium ion H_3O^+. The second is ammonium NH_4^+. Both of these can be important indicators of unusual soil conditions. The hydronium ion is associated with acidity in all media and is only present in appreciable amounts when the pH is acidic; that is, below 7. Because of its size and hydration it does not fit easily into any scheme designed to describe cations in soil.

Soil pH is rarely below 4.0, thus when pHs occur appreciably below 4 they indicate an unusual situation. Usually such low pHs are associated with some cold region forest soils or with sulfur and its oxidation to sulfuric acid. This is particularly likely in and around mining operations, in which sulfur is brought into oxidizing conditions, especially when it is mixed with soil or water. Whenever very low pHs are found in soil, the source should be found, because low environmental pHs indicate a serious problem that needs to be found and corrected. Very low pHs can solubilize otherwise inert toxic minerals and can ultimately lead to the total destruction of the soil itself.

The second common nonmetal cation is ammonium NH_4^+. Ammonium is important because plants use it as a source of nitrogen for plant growth and development. Ammonia is also important because it can be oxidized to both nitrite and nitrate.

Both H^+ and NH_4^+ act as cations in soil and become exchangeable cations on the cation exchange capacity. Although protons H^+, or H_3O^+ are cations, it is hard to define their exact cation exchange activity. Ammonium (NH_4^+), however, acts like any other cation in the soil and can readily be identified as an exchangeable cation. Both these cations can affect cation availability and exchangeability in soil analysis. Exposure of samples to either changes in pH or ammonia thus can change analytical results.

9.2.5. Cation Exchange Capacity

Clay and organic matter in soil usually have a negative charge and thus attract and hold cations. This results in the soil having cation exchange capacity (CEC). It should be expected that a portion of all cations in soil will be associated with the cation exchange sites. This includes both metal and nonmetal cations.

Analytical methods cannot directly measure the amount of cations on the exchange sites; they must be removed from the sites and brought into solution before analysis. This means that some replacing cation must be used to exchange with them and move them into solution. A cation that has higher oxidation states (2^+ and 3^+) or that is present at high concentration will replace cations with lower oxidation states or lower concentrations.

Some cations are not attracted to CEC sites in soil. Most iron (Fe^{3+}), aluminum, titanium, and manganese precipitate as oxides and thus are not available to act as soluble or exchangeable cations. On the other hand, Fe^{2+} forms soluble complexes with organic compounds, and thus it also does not act as a simple cation. It should be expected, however, that both oxidation states of iron will be found in most soils.

Although cation exchange capacity is most often discussed, some soils also show appreciable anion exchange capacity. Some tropical and highly weathered soils are particularly noted for this characteristic. Just as in cation exchange, an anion must be removed from anion exchange sites to be analyzed. It is also important to remember that not all anions (e.g., phosphate) are either excluded or attracted to soil particles in an exchangeable fashion [16].

9.2.6. Anions

Anions in soil are single oxidation state species, such as the halogens and the multioxidation state oxyanions. Only the halogens are commonly found in the environment as simple anions and are found with an oxidation state of negative one. Because most soil particles and the simple anions have negative charges they repel each other. Thus the simple anions are readily leached lower in the soil or out of it altogether [17]. The oxyanions may or may not be readily leached out of soil, depending on their ability to chemical react with other ions or species.

Many oxyanions occur naturally in the environment. The most important of these are carbonate, nitrate, phosphate, and sulfate. There are many others, however, that are in lower concentration but are nevertheless extremely important. Important oxyanions, their structures, and their biological importance are given in Table 9.2.

9.2.6.1. Carbonate and Bicarbonate

In addition to their biological importance, carbonate and bicarbonate are important in determining the buffering and pH of soil and soil solutions. Carbon dioxide from respiration in plant roots and microorganisms is released into the soil atmosphere. Some dissolves in water, forming bicarbonate ions.

$$CO_2 + H_2O \longrightarrow HCO_3^- + H^+$$

If the existing soil solution becomes acidic, bicarbonate will accept a proton-

TABLE 9.2 Common Environmental Oxyanions

Name	Formula	Source and biological importance
Carbonate	CO_3^{2-}	Counter and exchange ion in root uptake of cations and anions
Bicarbonate	HCO_3^-	Counter and exchange ion in root uptake of cations and anions
Sulfate	SO_4^{2-}	Source of sulfur for plants
Nitrate	NO_3^-	Source of nitrogen for plants
Nitrite	NO_2^-	Source of nitrate in soil
Phosphate	PO_4^{3-}	Source of phosphorus for plants
Molybdenate	MoO_4^{2-}	Source of molybdenum for plants and microorganisms; important in nitrogen fixation
Arsenate	AsO_4^{3-}	Toxic component in the environment
Selenate	SeO_4^{2-}	Source of selenium for plants and animals
Borate	BO_3^{3-}	Source of boron for plants

producing carbon dioxide and water, the reverse of the equation above. The common test for carbonate minerals is

$$CaCO_3 + 2H^+\, 2Cl^- \longrightarrow CO_2 + Ca^{2+} + 2Cl^- + H_2O$$

to treat them with dilute hydrochloric acid. The release of gas in the form of bubbles is a positive test for carbonate. Soil carbonates form in soil by the following reaction:

$$2H_2O + CO_2 + Ca^{2+} \longrightarrow CaCO_3\downarrow + 2H^+$$

In basic environments protons are released from bicarbonate and carbonate is formed as shown above. If this takes place where there is an excess of calcium or some other base cation, a carbonate such as calcium carbonate is formed. Carbonates are not very soluble in water, and this reaction results in deposition or precipitation.

Because of its involvement in maintaining or controlling pH, carbonate is an important oxyanion. Exposure of samples to carbon dioxide may affect the pH of the sample and consequently the availability and solubility of many components in that sample. Exposure to high levels of carbon dioxide during sampling or sample transport is thus expected to have an effect on the analytical results.

9.2.6.2. Sulfur and Sulfate

Sulfur can occur as an element in nature. In soil and water, however, it does not remain in the elemental form long. Under reducing conditions it is reduced to hydrogen sulfide (H_2S), and under oxidizing conditions it is oxidized to sulfuric acid (H_2SO_4). Soils and soil samples exposed to elemental sulfur, most commonly labeled as flowers of sulfur, will suffer a dramatic decrease in their pH. This in turn can lead to the release of materials into the soil solutions that are not normally found there. Soil sulfur, however, in the form of sulfate as the anion (SO_4^{2-}), is always present, is stable, and does not affect pH.

Sulfur is an essential element in the nutrition of plants, animals, and microorganisms. Plants can take in many different forms of sulfur—sulfate being one of them—and incorporate it into various needed organic sulfur compounds. Calcium and magnesium sulfates are commonly used to ameliorate various soil conditions caused by poor soil structure and are also used as a source of calcium under conditions in which it is not desirable to change a soil's pH [17].

9.2.6.3. Phosphate

The oxyanion phosphate (PO_4^{3-}) is a common and essential component of life. It occurs in different forms, depending on the pH of its surroundings and associated metals. Commonly one or two of its negative charges are satisfied by protons, and thus it occurs as $H_2PO_4^-$ or as HPO_4^{2-}. These are the two most common forms occurring in the normal pH range of water and soil. The 1^- and 2^- charges on these forms are typically satisfied by sodium, potassium, calcium, or magnesium.

Under very acid conditions phosphate occurs as phosphoric acid (H_3PO_4), while under very basic conditions it occurs as phosphate (PO_4^{3-}). Because of the extreme pHs under which these forms are found, their occurrence indicates unusual soil or environmental conditions.

Looking at phosphate it would be assumed that this oxyanion would act as all anions are expected to act. It does not. All species of phosphates react with metals to form insoluble, plant-unavailable phosphates. Under acid conditions either monobasic ($H_2PO_4^-$) or dibasic phosphate (HPO_4^{2-}) reacts with iron and aluminum to form-insoluble and unavailable iron and aluminum phosphates. Under basic conditions insoluble and unavailable calcium phosphates form. Phosphate species can also react with both solid inorganic surfaces and organic components, and thus are also removed from solution by these mechanisms. For these reasons phosphate does not typically move in soil unless unusual conditions, such as extremely high levels of phosphate or very sandy soils, occur [18].

High levels of plant-available phosphate in soil and water can occur naturally. In Florida, where phosphate is mined and soils are sandy, soil phosphate can be so high that it readily leaches into water. In any area in which phosphate is mined the Earth's surface may have high concentrations of phosphate. Both of these cases are easily explained. Such situations are unusual, however, and any time high levels of phosphate are found the source of phosphate needs to be determined.

9.2.6.4. Oxides of Nitrogen

The two most common oxides of nitrogen are the oxyanions nitrite and nitrate. Table 9.3 provides a list of all the common environmental oxides of nitrogen. Nitrogen is added in large amounts to agricultural land as nitrogen fertilizer, anhydrous ammonia, aqueous ammonia solutions, and various nitrates, including ammonium nitrate. Ammonium added to the soil when temperatures are above $5\,°C$ and in the presence of moist but not saturated water conditions is quickly oxidized to nitrite and then nitrate. Nitrite is usually present in small amounts because oxidation of nitrite to nitrate is faster than the oxidation of ammonia to nitrite. During the oxidation of ammonia to nitrate protons are released into the surroundings. Thus both soil and water may become acidic during this transformation.

Both nitrite and nitrate carry a negative charge. They are mobile in soil, and their compounds are very soluble. When either nitrite or nitrate are present in soil they can leach into the ground water. However this only occurs as long as they are under oxidizing conditions. Under reducing conditions nitrate is reduced to N_2, NO, and N_2O gases. The occurrence of nitrate along with the various oxides of nitrogen in environmental samples should be expected. When they are found, increased acidity may also be found. The occurrence of large amounts of nitrite in environmental samples indicates an unusual situation that needs further investigation [19].

TABLE 9.3 Common Environmental Oxides of Nitrogen

Name	Chemical formula	Common occurrence
Nitric oxide	NO	During denitrification
Nitrous oxide	N_2O	During denitrification
Nitrogen dioxide	NO_2	During denitrification
Nitrite	NO_2^-	First oxidation product of ammonia
Nitrate	NO_3^-	Final product in oxidation of nitrite

9.2.6.5. Molybdate

Molybdate is extremely important in biological nitrogen fixation. Molybdate is present in relatively low concentrations in soil, however. Acid soils with appreciable leaching may have total molybdate levels around 0.5 mg/kg. Basic soils having low rainfall may have levels 10 times that amount. Soils to which sewage sludge has been added may have significant increases in molybdate concentration. Because of its negative charge molybdate would be expected to move readily through the environment. It reacts with many constituents, however, and thus does not move as readily as expected [20].

9.2.6.6. Borate

Boron is present in soil and is taken up by plants as H_3BO_3. It can also occur as borate and $H_4BO_4^-$, however. The boron oxyanions are generally present in very low concentrations in soil. Boron can move by both mass flow and diffusion in the soil solution. Because it is both an essential nutrient for and is toxic to plants it is of particular environmental concern. There is a very narrow range over which it is sufficient but not toxic. High levels of boron in soil can result from coal ash disposal and represent a cause for concern when revegetation and bioremediation are undertaken [20].

9.3. THE ORGANIC COMPONENT

Organic compounds in the environment range from the very simple to the very complex. From methane, which is released during anaerobic decomposition of organic matter, to humus, which is the organic matter that is produced during organic matter decomposition and remains afterwards. Compounds of intermediate complexity are also formed during the organic matter decomposition process. All organic functional groups can be found in the organic components in the environment, and all of these are in addition to inorganic elemental carbon, carbon dioxide, and carbonate.

To be an organic compound a molecule must be made up of carbon and hydrogen. It may also contain several other different kinds of atoms, the most important of which are oxygen and nitrogen. Other atoms, however, are common, such as sulfur, phosphorus, and the halogens, particularly chlorine and bromine. When these other atoms are present, their position and arrangement in the molecule are extremely important. They form the functional groups that control a compound's interaction with the environment. Solubility in water, movement, reactions, and toxicity are all related to a compound's functional groups.

Halogens substituted on any organic compound have a dramatic effect on both their physical and chemical properties. Halogens are heavy, and substituting a halogen for a hydrogen on an organic compound increases the compound's density, sometimes making it denser than water. Halogens also make the compounds less soluble in water. On the other hand, many oxygen- and nitrogen-containing functional groups increase a compound's solubility in water.

9.3.1. Hydrocarbons

The backbone of all organic compounds is carbon atoms bonded together. The most highly reduced compounds are alkanes, which contain only carbon and hydrogen and are part of a group of compounds called hydrocarbons. As the name implies, these compounds contain only carbon and hydrogen. This group is composed of three families of compounds that are used as fuels, are common in the environment, and are common pollutants. The three families are the alkanes, alkenes, and alkynes. Alkenes have at least one double bond, and alkynes have at least one triple bond in addition to carbon, hydrogen and single bonds.

The hydrocarbons are less dense than water and are insoluble in water. Because of these characteristics they move with water through the environment by floating on it.

When analyzing soil for hydrocarbon contamination, it is common to treat any combination of hydrocarbons as a single unit or entity. In analytical results they are reported as the total amount of all of these types of compounds, and thus they are referred to as total petroleum hydrocarbons (TPH). This analysis thus gives no indication of the types of compounds present or their relative concentration. Indeed, a gas chromatographic analysis of hydrocarbon fuels gives a very complex and poorly resolved chromatogram with a large number of peaks showing the presence of a large number of compounds.

Groups that contain only carbon and hydrogen are called alkyl groups and are represented by the letter R. This represents a group of any configuration that contains only carbon and hydrogen and is attached to another, usually larger, component. In some cases an R group may be used to represent other constituents in a molecule. If this is the case, however, this usage must be expressly and distinctly shown.

In addition to compounds that contain only carbon and hydrogen are numerous environmentally sensitive compounds that contain carbon, hydrogen, and one or more halogen atoms. The halogens are substituted for hydrogens in these compounds and are generally called halocarbons. They are similar to hydrocarbons in that they are nonpolar, insoluble in water, sometimes denser than water, and generally slowly biodegradable. When denser than water halocarbons will be at the bottom rather than at the top of aqueous environments and aqueous extracts of soil.

Most halocarbons are man-made or the result of man's activities, some examples of which are given in Table 9.4. There are, however, a large number (over 1,500), of natural halogenated organic compounds occurring in the environment. A small amount of chlorinated organic compounds are produced by such common soil microorganisms as the bacteria *Pseudomo-*

TABLE 9.4 Common Environmentally Important Halogen-Containing Organic Compounds

Name	Structure
Chloromethane	$ClCH_3$
Dichloromethane	Cl_2CH_2
Trichloromethane (chloroform)	Cl_3CH
Tetrachloromethane (carbon tetrachloride)	Cl_4C
Dichlorodifluoromethane (a CFC)[a]	Cl_2CF_2
1,2-Dichloroethane	$H-\overset{\displaystyle H}{\underset{\displaystyle Cl}{C}}-\overset{\displaystyle H}{\underset{\displaystyle H}{C}}-Cl$
1,2-Dichloroethene	$\underset{Cl}{\overset{H}{}}C=C\underset{H}{\overset{Cl}{}}$
Freons[a] (used as refrigerants)	$Cl-\overset{\displaystyle F}{\underset{\displaystyle Cl}{C}}-F$
Trichloroethylene	$\underset{H}{\overset{Cl}{}}C=C\underset{Cl}{\overset{Cl}{}}$
Vinyl chloride (monomer used in the production of plastics)	$\underset{H}{\overset{H}{}}C=C\underset{Cl}{\overset{H}{}}$

[a]Numbers of chlorines and fluorines varies in chlorofluorocarbons (CFCs).
Note: Most of these compounds are used as solvents.

nas and *Bacillus*, the actinomycetes *Streptomyces* and *Saccarothrix*, and the fungi *Aspergillus* and *Penicillium*. It thus cannot be assumed that all halogenated organic compounds are anthropogenic, nor can it be assumed that all halogenated organic compounds are resistant to degradation in the environment [21].

9.3.2. Aromatic Compounds

Another group of compounds with the characteristics of hydrocarbons are those containing benzene rings. Another name for these hydrocarbons is aromatic, although they often do not have a sweet or spicy fragrance. Many aromatic compounds are of environmental concern as toxins or carcinogens; however, an equal or greater number of compounds containing benzene rings are inert or even beneficial. Tryosine, an amino acid, cinnamonaldehyde, a component of cinnamon responsible for its smell and taste, and aspirin, an analgesic, are but three benzene ring-containing compounds regularly used by humans. Phenylalanine, another amino acid that contains a benzene ring, is an essential dietary amino acid.

Compounds containing a benzene ring can be classified as belonging to one of three groups. A relatively simple group is those compounds that contain only one benzene ring having some group attached to it. Examples are the benzene itself, along with toluene, ethylbenzene, and the xylenes. This group of compounds is often referred to as BTEX, meaning some combination of all of these compounds. Any benzene ring with an alkyl group may also be referred to as an alkylbenzene. Compounds of this type are shown at the top of Figure 9.3. Hydrogen atoms are at each corner of the hexagon rings, as shown for benzene C_6H_6, but they are omitted in the other diagrams for simplicity.

A second group of compounds is made up of benzene rings bonded to each other by a single bond or where two or more rings share carbons. An example of the first type of compound is biphenyl, which has two benzene rings joined by a single bond. Compounds containing only the benzene rings carbon hydrogen and halogens can be either toxic or carcinogenic. One important group of these compounds is the polychlorinated biphenyls (PCBs). An example of a polychlorinated biphenyl is shown in Figure 9.3. There are 10 carbons that can have a chlorine substituted for hydrogen on either or both of the two rings. There are thus a large number of possible PCBs.

Examples of the second type of compound are situations in which the benzene rings share two carbons (e.g., naphthalene or anthracene). These condensed ring compounds may have the rings joined in many different ways, as shown in Figure 9.3, producing different compounds. The

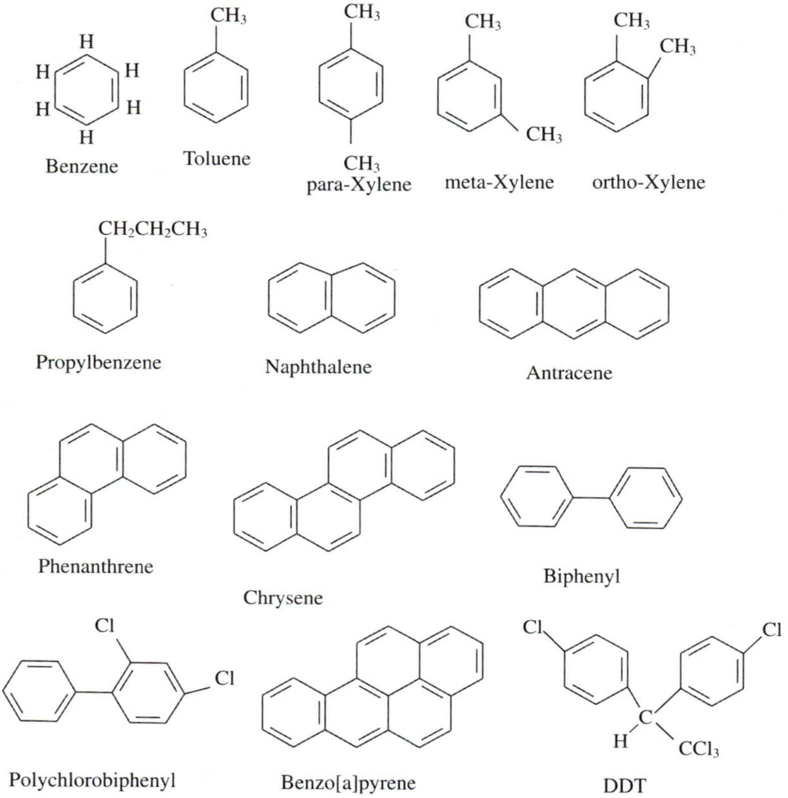

FIGURE 9.3 Common aromatic compounds containing one and several benzene rings bonded together. Hydrogen atoms are at the corners of the hexagons, as shown for benzene, but are omitted in the others for simplicity. DDT is dichloro-diphenyltrichloroethane.

benzo[α]pyrenes are examples of a type of compound sometimes called a "chicken wire" compound. Many of these compounds, such as ben-zo[α]pyrene itself, are carcinogenic and thus are of environmental concern.

A third group is compounds containing two or more benzene rings separated by nonbenzene groups. This includes a large number of compounds that contain benzene rings separated by a variety of substituted and unsubstituted alkyl groups. The connecting groups thus may have other functional groups attached to them. Of particular concern are those types of compounds containing halogens. Some of these, such as DDT, are of general environmental concern; others are of particular human health concern. Two chlorinated aromatic compounds of this type are given later in Figure 9.5.

9.3.3. Oxidized Functional Groups

The partially oxidized organic compounds contain acid, aldehyde, ketone, ether, and alcohol functional groups. (See Table 9.5.) Each functional group makes up a family of compounds having different reactivity, solubility, and interaction with various constituents in the environment. They are generally more soluble in water and move more readily in the environment than do hydrocarbons. In many cases they are also more easily decomposed by microorganisms to CO_2 and water.

Organic molecules can contain more than one functional group. Common examples are sugars (hexoses), which contain five alcohol (–OH) groups, and one aldehyde (–CH $=$ O) group, as seen in Figure 9.4. Another common example is unsaturated fatty acids, which contain one acid functional group and one or more double bonds. Amino acids contain at minimum one acid functional group and one amine group.

Increasing the number of oxygen- or nitrogen-containing functional groups in a compound increases its solubility. In Figure 9.4 six carbon hexane is very insoluble in water, while the six carbon hexose sugar is very soluble. Sugars are polymerized into starch and cellulose, fatty acids polymerize into fats, and amino acids polymerize into proteins, and when this happens the size of the polymer and its linkages will control solubility rather than numerous oxygen- or nitrogen-containing functional groups.

When larger organic compounds degrade in the environment, they undergo a series of decomposition steps during which a number of different intermediate organic compounds are produced. These often contain a number of different functional groups. The rate of decomposition of starting and intermediate compounds is related to the functional groups they contain. Functional groups control a compound's solubility in water. They also control its rate of decomposition, because generally, the more water-soluble a compound is the faster it is degraded. The more soluble a compound is, however, the more likely it is to move in the environment, and thus there will be an increased need for sampling.

This also means that when looking for a particular organic contaminant in soil or water its natural occurrence must be established and quantified. It cannot be assumed that the compound of interest is not naturally or commonly present in a particular environment. This even means that highly reduced compounds such as alkanes, alkenes, and alkynes may be present even though their formation in aerobic environments is

TABLE 9.5 Organic Functional Groups

Group	Structure	Characteristics
Alkanes	$CH_3(CH_2)nCH_3$	Contains only carbon and hydrogen. Gases and liquids insoluble in water, flammable.
Alkenes	$CH_3 — CH = CH_2$ Propene	Contains at least one double bond. Gases and liquids insoluble in water, flammable.
Alkynes	$CH \equiv CH$ Ethyne	Contains at least one triple bond. Gases and liquids insoluble in water, flammable.
Alcohols	CH_3CH_2OH Ethanol	Contains at least one —OH group. Low molecular weight alcohols are soluble in water, liquids and solids.
Ethers	$CH_3CH_2 — O — CH_2CH_3$ Diethyl ether	Contains at least one R—O—R group. Gases and liquids insoluble in water and flammable.
Aldehydes	$CH_3C\overset{\displaystyle O}{=}\underset{H}{}$ Ethanal	Contains at least one $R—\overset{\displaystyle O}{\underset{\displaystyle \|}{C}}—H$ group, liquids and solids.
Ketones	$CH_3C\overset{\displaystyle O}{=}CH_3$ Propanone	Contains at least one $\overset{R}{\underset{R}{}}C=O$ group, liquids and solids.
Acids	$CH_3C\overset{\displaystyle O}{=}OH$ Acetic acid	Contains at least one $R—\overset{\displaystyle O}{\underset{\displaystyle \|}{C}}—OH$ group, liquids and solids.
Amines	$CH_3CH_2NH_2$ Ethyl amine	Contains at least one $(R)_3—N$ group, liquids and solids.

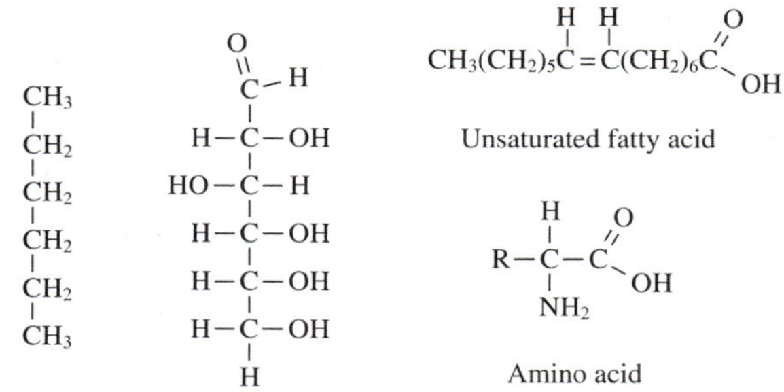

Hexane   Sugar   Unsaturated fatty acid   Amino acid

FIGURE 9.4 Structure of some common multifunctional organic compounds and hexane.

unlikely. On the other hand, partially oxidized compounds are also possible. Two examples not naturally found in the environment are 2,3,5-trichlorophenoxyacetic acid (2,4,5-T), an herbicide, and 2,3,7,8-tetrachlorobenzo-p-dioxin. The structures of these compounds are given in Figure 9.5.

A number of abbreviations and acronyms are used in discussing organic compounds in soil. BTEX and PCB have already been discussed. In addition, all volatile organic compounds will be grouped together and called VOCs, through this term usually refers to environmentally hazardous volatile compounds. Compounds that do not dissolve in water can be called nonaqueous phase liquid (NAPL). Aromatic compounds containing two or more benzene rings will be referred to as polyaromatic hydrocarbons (PAH). Other abbreviations and acronyms also are in common use. (See Chapter 1 and Appendix A.)

9.3.4. Humus

As stated above, during the process of microbial decomposition new organic matter is synthesized. This results in the formation of humus, which is a

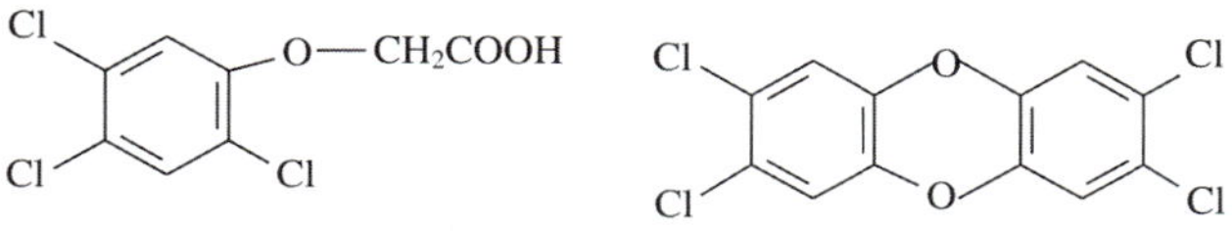

2,4,5-T   Dioxin

FIGURE 9.5 Examples of compounds containing a variety of alkyl, benzene, and functional groups. Hydrogen atoms are omitted for simplicity.

complex compound composed of a large number of diverse groups. Some have said that humus is a polymer. No monomer or mer* had ever been identified, however, and so it cannot really be said to be a true polymer. It is a large, complex molecule, that is dark in color and resistant to decomposition.

Soil scientists and others routinely separate humus into a number of different components: humic acid is extracted with alkali, fulvic acid is soluble in acid, and humin is not soluble in any solvents. Being able to separate humus into different components has not generally led to an illumination of its structure. Knowing the structure of these separate parts also does not shed much light on the chemical characteristics of humus, and humus as a whole has characteristics that are different from those of any one of these components. In spite of not knowing its structure it is known that humus has a pronounced effect on water and soil and their characteristics. This is particularly true when trying to describe humus and soil interaction with environmental pollutants.

Because of its complex organic structure and numerous functional groups, humus has a high affinity for many organic compounds and metals. (See chelation below.) It does not act like it has a surface, but as if it has organic compounds dissolved in it. In some cases it may be hard to tell when the dissolved contaminant ends and the humus begins. Humus will hold many different types of organic compounds containing many different types of functional groups. Once adsorbed by humus, compounds are generally slower to move in soil.

Colloidal organic matter and humus may remain suspended in water, sorb components, and move through the environment. This may result in movement of components not expected to move and result in their being present in higher than expected concentrations. Humus has a strong affinity for metals because of its cation exchange sites, and so it will attract and hold cations by cation exchange mechanisms. The two groups most often sited as being responsible for cation exchange in humus are acid and phenolic groups. Phenolic groups are benzene rings with an –OH group. Both acid and phenol groups can ionize, leaving a negative charge on the group and thus attracting cations.

Another mechanism of attraction is through its oxygen-containing groups, by which humus can hold metal cations in structures resembling chelates, as shown in Figure 9.6. (Chela is Latin for crab.) In this case association with lone pairs of electrons on oxygen and nitrogen hold the metal. The term chelate indicates that there are several parts of one organic

---

* A mer is the individual unit or units of which a polymer is made.

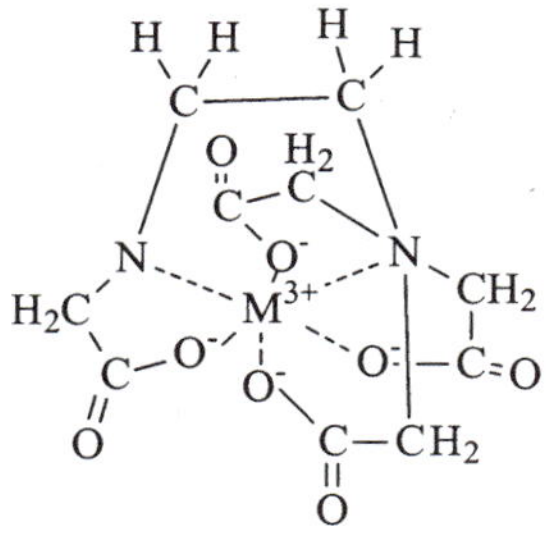

FIGURE 9.6 Ethylenediaminetetraacetic acid with a chelated metal ion (M+).

compound involved in the association. An example of a simple organic chelate is ethylenediaminetetraacetic acid (EDTA). In this case the functional groups act like claws associated with the metal. Humus also has a high affinity for metals and holds them by both cation exchange and chelation mechanisms. These mechanisms hold metals in forms that are available to plants [22,23].

9.4. THE BIOCHEMICAL COMPONENT

In addition to simple and complex organic compounds, soil also contains compounds that are biochemical. These occur from two different sources. First, many organisms exude enzymes to decompose compounds outside the cell. In most if not all cases this is done so that molecules that are too large to bring inside the cell for digestion can be broken down. For instance, cells cannot use cellulose until it is broken down into its individual sugar monomers (e.g., fungi), which are the primary cellulose degraders and exude extracellular enzymes, cellulases, to break down cellulose.

Extracellular enzymes are most often discussed as being exuded by microorganisms; however, there is evidence that they are also exuded by plants and even some animals. Plant roots have been shown to exude the enzymes invertase, phosphoratase, and amylase. Animals, particularly earthworms, have been shown to contribute invertase to the soil solution. This is similar to our secreting invertase in saliva to break down starch to sugar in our mouths. Little is known about the overall importance of animal-released enzymes in soil, however.

Extracellular enzymes are of particular interest because of their involvement in soil chemical reactions, and consequently they have been extensively studied. They can catalyze numerous reactions and result in both decomposition and intermediate compounds being present. Extracellular enzymes, which decompose proteins, fats, hydrocarbons, and cellulose, should be expected to occur in any soil that has these components in it. Some common examples of enzymes found in soil and their source are given in Table 9.6.

Enzymes are often associated with the mineral fraction in soil. The association of enzymes with clay minerals in particular has been extensively investigated. Often such associations are more active catalytically than are the enzymes alone, thus even when the enzymes do not occur freely in the soil solution they may still be present and active [24]. Some inorganic and even some organic components in soil may catalyze the same reactions as enzymes, however. Peroxide decomposition and hydrolysis can be promoted by inorganic ions, exchange sites, and organic molecules at normal soil temperatures [24], thus the observation of a reaction occurring in soil may not always be clearly related to enzymes alone.

Other compounds may also be exuded into the environment by organisms. Some may be simple organic waste products and others complex molecules that the organism cannot use. These may include intermediate decomposition products from the breakdown of pesticides and insecticides.

Microbial, plant, and animal cells in or added to the soil both exude material and break open, releasing their contents. Any intercellular component may thus be found in soil at any time. Many of these compounds are easily decomposed by other organisms and so do not last long in the soil environment. Some compounds may become associated with clay particles and become a little more resistant to further decay. In the decomposition process many intermediates are found and may be part of a mix of compounds found in a soil analysis.

Cellular components that are not easily decomposed will also be present. These may include both organic compounds and inorganic compounds. Microorganisms can reduce sulfate, nitrate, selenium, and

TABLE 9.6   Some Common Extracellular Enzymes and Their Common Source

Enzyme	Source
Proteases, cellulases, and pectic and proteolytic enzymes	Numerous bacteria and fungi
Cellulase	Fungi
Lipases	Fungi and other microorganisms
Phosphatase	*Bacillus, Fusarium, Saccharomyces*, and many other soil bacteria and fungi

Source: Taken from Ref. 24.

other elements to their elemental form, and these may thus be present in soil from this source. They may also be present in combined form as a result of microbial activity. Cell walls, compounds with long hydrocarbon moieties such as lignin, and other structures resistant to decomposition may persist in the soil for significant amounts of time [25].

9.5. THE LIVING COMPONENT

All soil has organisms living in it, including animals, plants, and microorganisms. The number of microbes varies with the counting method, and is as high as 10^9 per g of soil. The activities of these organisms will affect the sampling plan and analytical procedures. One affect is soil component decomposition, and another is their movement through soil. Such organisms as ants and termites can move large amounts of soil. Often this means moving soil from some depth to the soil surface. This is important because it might be assumed that the A horizon is being sampled when in reality it is the C or B horizon that is really being sampled and analyzed.

In some cases the reverse may be true. Groundhogs dig sizable tunnels down and through the soil profile. When the animals abandon these holes they may fill with soil washed into them from the surface. In this case samples assumed to be from the B or lower horizons may actually be from the surface. This may not be evident by observing a single sample, but would be obvious when observing a soil profile. The same type of thing can happen with other animals and animal holes in soil.

Where a spill has happened in this type of environment, contaminants can move deep into the soil through these types of holes. In this case sampling must be deeper than the holes if an accurate determination of the extent of contamination is to be known. In both of these cases knowing that borrowing animals are common and that burrows have been filled in is essential. Such information is available in a soil survey.

The above examples are for groundhogs, ants, and termites; however, in different parts of the world different animals will be living in soil and making it their home. It should not be assumed that because groundhogs do not live in a particular area this type of soil mixing cannot happen. It can.

Plants change the soil in a more dramatic way. As roots grow through the soil they change its physical and chemical properties. Roots push soil particles together, allowing them to be held together by cations, clay, organic matter, and microbial gums to form secondary particles, or peds. Secondary particles have distinct areas of weakness and void volumes between them that increase infiltration and percolation. This in turn will increase contaminant movement through the soil. Sampling will need to take these changes into consideration.

Roots exude acids and give off carbon dioxide. Both of these make the soil more acid. The roots also deplete the soil around them of water and base cations. As it moves toward the roots, water can carry components, including contaminants, to plants that either collect at the root surface or are taken into the roots and transported to the plant leaves. Soils with a dense plant cover such as sod* will have characteristics that are different from soils that have row crops or soils which are sparsely vegetated.

Microorganisms can also change soil's physical and chemical characteristics. Through the production of microbial gums, sand, silt, and clay particles can be cemented together and change infiltration and percolation through soil. More important, microorganisms can dramatically change the chemical characteristics of soil. Rapid decomposition of organic matter can deplete soil air of oxygen, creating anaerobic conditions. Microorganisms can also change the oxidation state of inorganic ions around them. In this way microorganisms can dramatically change the components present and therefore the components one is looking for may be dramatically changed. Final and intermediate decomposition products may be present in the same sample.

Because of these differences sampling will be different for different environments. Typically sod can be sampled on a random basis. Pastures with animals will be a little different because of animal tracks and manure, both of which should be avoided when sampling. In tree-covered areas there may or may not be vegetation between the trees. In any case sampling must be done between trees as randomly as possible. In industrial areas the same concepts must be observed; that is, plants may accumulate contaminants. Animal holes and other soil cracks will allow components to move rapidly into and through soil; thus all these potential sampling considerations need to be kept in mind.

In addition to animals being present their activity is also important. Any environmental condition that changes the activity of animals may also change the occurrence of the compounds associated with them. Unusually wet, dry, hot, or cool conditions and the seasons will change animal and plant activity, thus weather conditions and the date can change the types of compounds found, and this should be noted in the project notebook and taken into consideration when interpreting analytical results [26].

---

* Sod refers to any situation in which soil has a continuous grass cover (e.g. lawns, golf courses, hay fields).

9.6. CONCLUSIONS

Soil contains a large and diverse number of inorganic, organic, and biological molecules. The inorganic component comes from the breakdown and leaching of rock and minerals. The organic components come from organic matter added to the soil by plants, animals, and microorganisms. This same group provides biochemical molecules, enzymes, proteins, fats, carbohydrates, nucleic acids, and lignins to the soil. In addition, the breakdown of animal and plant tissue releases intracellular constituents to the environment, particularly the soil solution. Because of the intimate relationship between soil and water, these compounds may be found in water. Any of these compounds that are gaseous can also be found in the atmosphere. In addition to chemical change plants, animals and microorganisms can create physical changes in soil that affect sampling.

It would simplify things greatly if all the reactions taking place in soil could be elucidated and described fully. There are always new and often unexpected reactions being discovered, however. For instance, microorganisms can carry out both transhalogenation and chlorine bromine exchange in halomethane. This complicates both the quantification and the understanding of the source and fate of these compounds in the environment [27].

QUESTIONS

1. List the 10 most common elements and inorganic compounds found in soil.
2. What are the two nonmetal cations found in soil, and why are they particularly important?
3. Phosphate is an oxyanion, but it does not move in soil. Explain why it does not move.
4. All organic compounds are composed of what two elements?
5. Organic molecules made up of only carbon and hydrogen are known by what general name?
6. Compounds containing benzene rings are of particular concern in the environment. Give examples of three groups of compounds containing benzene rings. Why are these groups of particular environmental concern?
7. Draw the structures for four common oxidized functional groups found in organic compounds. These functional groups result in the compounds containing them having increased water solubility. Explain how this happens.
8. One group of extremely important biochemical component in soil are enzymes. What are enzymes and how do they function in soil?

9. Explain how animals change soil and how they can affect sampling plans.

REFERENCES

1. Brady NC, Weil RR. The Nature and Properties of Soils. 12th ed. Upper Saddle River, NJ: Prentice-Hall, 1999:268–269.
2. Sato M, Sutton AJ, McGee KA, Russell-Robinson S. Monitoring of hydrogen along the San Andreas and Calaveras Faults in central California in 1980–1984. J Geophys Res Gas Geochem Volcan Earthquakes Resource Explor Earth Inter 1986; 91:12315–12326.
3. Virk HS, Walia V, Kumar N. Helium/radon precursory anomalies of Chamoli Earthquake, Garhwal Himalaya, India. J Geodyn 2001; 31:201–210.
4. Poissant L, Casimir A. Water–air and soil–air exchange rate of total gaseous mercury measured at background sites. Atmosph Environ 1998; 32:883–893.
5. Glaser B, Haumaier L, Guggenberger G, Zech W. Black carbon in soils: The use of benzenecarboxylic acids as specific markers. Org Geochem 1998; 29:811–819.
6. Sergeev NB, Gray DJ, Scott KM. Gold mass balance in the regolith, Mystery Zone, Mt. Percy, Kalgoorlie, Western Australia. Geochem Explor Environ Anal 2001; 1:307–312.
7. Alma MGM, Tokunaga S, Maekawa T. Extraction of arsenic in the synthetic arsenic-contaminated soil using phosphate. Chemosphere 2001; 43:1035–1041.
8. Papachristodoulou CA, Assimakopoulos PA, Patronis NE, Ioannides KG. Use of HPGe γ-ray spectrometry to assess the isotopic composition of uranium in soils. J Environ Radioact 2003; 64:195–203.
9. Raju KK, Raju AN. Biogeochemical investigation in south eastern Andhra Pradesh: The distribution of rare earths, thorium and uranium in plants and soils. Environ Geol 2000; 39:1102–1106.
10. Konrad KB, Introduction to Geochemistry. New York: McGraw-Hill, 1967:99–105.
11. Strojan ST, Philips CJC. The detection and avoidance of lead-contaminated herbage by dairy cows. J Dairy Sci 2002; 85:3045–3053.
12. Foth H. Fundamentals of Soil Science. 8th ed. New York: Wiley, 1990:14.
13. Shuman LM, Chelate and pH effects on aluminum determined by differential pulse polarography and plant root bioassay. J Environ Sci Health. pt A: Environ Sci Eng Toxic Hazard Subst. Control 1994; 29:1423–1438.
14. Nico PS, Anastasio C, Zasoski RJ, Rapid photo-oxidation of Mn(II) mediated by humic substances. Geochimica et Cosmochimica Acta 2002; 66:4047–4056.
15. Campillo N, Lopez-Garcia I, Vinas P, Arnau-Jerez I, Hernandez-Cordoba M. Determination of vanadium, molybdenum and chromium in soils, sediments and sludges by electrothermal atomic absorption spectroscopy with slurry sample introduction. J Anal Atomic Spectrom 2002; 17:1429–1433.

16. Becquer T, Pétard J, Duwig C, Bourbon E, Moreau R, Herbillon AJ. Mineralogical, chemical and charge properties of geric ferralsols from New Caledonia. Geoderma 2001; 103:291–306.
17. Havlin JL, Tisdale SL, Nelson WL, Beaton JD. Soil Fertility and Fertilizers: An Introduction to Nutrient Management. 6th ed. Upper Saddle River, NJ: Prentice-Hall, 1999:222–230.
18. Donahue RL, Miller RW, Shickluna JC, Soils: An Introduction to Soil and Plant Growth. 4th ed. Englewood Cliffs, NJ: Prentice-Hall, 1977:131–133.
19. Brady NC, Weil RR, The Nature and Properties of Soils. 12th ed. Upper Saddle River, NJ: Prentice-Hall, 1999:495–519.
20. Havlin JL, Tisdale SL, Nelson WL, Beaton JD, Soil Fertility and Fertilizers: An Introduction to Nutrient Management. 6th ed. Upper Saddle River, NJ: Prentice-Hall, 1999:276–281, 286–291.
21. Öberg G. The natural chlorine cycle—Fitting the scattered pieces. Appl Microbio Biotech 2002; 58:565–581.
22. Swift RS, Organic matter characterization. In: Bartels JM, ed. Methods of Soil Analysis: Part 3—Chemical Methods. Madison, WI: Soil Science Society of America, and Agronomy Society of America, 1996:1011–1069.
23. Sawhney BL. Extraction of organic chemicals. In: Bartels JM, ed. Methods of Soil Analysis: Part 3—Chemical Methods. Madison, WI: Soil Science Society of America and Agronomy Society of America, 1996:1071–1084.
24. Skujins JJ. Enzymes in soil. In: McLaren AD, Peterson GH, eds. Soil Biochemistry. New York: Marcel Dekker, 1967:371–414.
25. Soil Biochemistry. McLaren AD, Peterson GH, eds. New York: Marcel Dekker, 1967:371–414.
26. Brady NC, Weil RR, The Nature and Properties of Soils. 12th ed. Upper Saddle River, Prentice-Hall NJ: 1999:404–444.
27. Harper DB, Kalin RM, Larkin MJ, Hamilton JTG, Coulter C. Microbial transhalogenation: A complicating factor in determination of atmospheric chloro- and bromomethane budgets. Environ Sci Tech 2000; 34:2525–2527.

10

An Overview of Basic Principles of Analytical Methods

Rolf Meinholtz

The planning for an analytical strategy begins with the original question asked by the client and runs through sampling of the field in question, subsequent characterization of the sample, sample analysis, and data reduction to create the final report. During this process it is necessary to ask some essential questions at a fundamental level. The first is, What is the original question that is being asked by the client? Two basic factors must be considered here. The first is the amount and nature of the sample; that is, is it solid, liquid, or gas? The second is the potential concentration levels of components in the sample. Once these questions are considered and an analytical strategy is developed, it is time to take a sample. The entire process from initial question to final report is seen in the flow chart in Figure 10.1.

Once the sample has been taken some of it is enclosed in containers and sent to the commercial analytical laboratory for analysis. The rest is used to perform what are termed field parameters, field tests, or field analyses. Field analyses are those that—because the analytes change rapidly—must be measured immediately and directly [e.g., dissolved oxygen (DO)]. (See Tables 10.1 and 10.2 for examples of analyses done in the field, those done in the commercial laboratory, and hold times.) These are general lists, and in some cases additional field analyses may be dictated by the client or the analytical laboratory, depending on the analytical

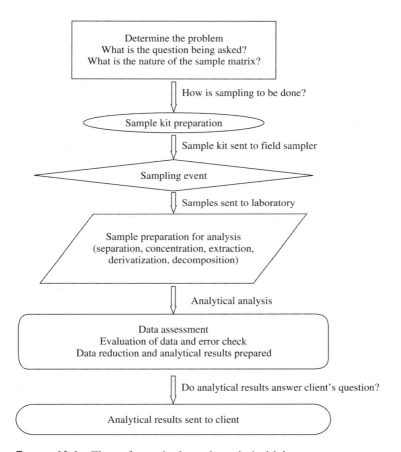

FIGURE 10.1 Flow of sample through analytical laboratory.

TABLE 10.1 Analytical Procedures Which Can Be Carried Out in the Field or Properly Equipped Field Laboratory

Color	Dissolved oxygen
Iodine	Odor
pH	Sulfite
Temperature	Total dissolved solids
Chlorine	

TABLE 10.2 Recommended Maximum Holding Times for Samples Between Sampling and Analysis[a]

Time from sampling	Analysis
Immediate	pH
6 hr	Fecal coliform
	Standard plate count
	Ground and wastewater total coliform
24 hr	502/524 (unpreserved)
	Hexavalent chromium
	Odor
	Residual chlorine
	THM and HAA formation potential
30 hr	Total coliform in DW
36 hr	Bioassay
48 hr	Asbestos
	BOD/CBOD
	Chlorophyll A
	Color
	Dissolved oxygen (modified Winkler method)
	Nitrate/nitrite
	Orthophosphate
	Surfactants
	Turbidity
	UV254
72 hr	EPA 18 (air)
	Radon
7 days	Most extractable organics
	Sulfide
	TDS/TSS
	VOC aromatics (unpreserved)

[a]See EPA web site http://epa.gov/ for specifics of analysis.
Note: Holding times start at the time of taking the sample!

measurements requested. Once the field analyses are performed and the data recorded on what are termed field calibration or data sheets (Figure 10.2), this sample is usually discarded. These data sheets must be submitted to the laboratory with the associated samples since these results can change the analytical procedures used, as well as having an effect on the analytical results obtained [1].

Once the sample has been received and logged into the testing laboratory, the next step is sample preparation. In rare cases what are termed direct methods—which allow sample analysis without prior

Field Calibration Sheet	Submission #:						
Sampler: _____							
Date: _____							
pH Calibration	Recheck calibration every 4 hr						
Time	Standard Value	Initial Cal.	Verification	Time	Initial Cal.	Verification	Verification
	4.00						
	7.00						
	10.00						
Meter _____				Serial # _____			

Conductivity Calibration							
Time	Standard Value	Initial Cal.	Verification	Time	Initial Cal.	Verification	Verification
	DI Water						
	0.718						
	6.67						
	58.7						
	147						
	1,413						
	12,880						
Meter _____				Serial # _____			

FIGURE 10.2 Field calibration form.

preparation—might be used. However, for the majority of samples some preparation will be necessary before analysis.

Sample preparation can be considered to be the removal of interfering matrix components from the sample before analysis. It might involve modification of the sample using extraction, concentration, and/or other

Turbidity Calibration								
Time	Standard Value	Initial Cal.	Verification		Time	Initial Cal.	Verification	Verification
	DI Water							
	1							
	5							
	10							
	100							
Meter _____					Serial # _____			

Dissolved Oxygen Calibration								
Time	Standard Value	Initial Cal.	Verification		Time	Initial Cal.	Verification	Verification
	0.00							
	D.O.							
	Temp.							
Meter _____					Serial # _____			

FIGURE 10.2 Continued.

steps that change the original or "raw" sample into a form that can be analyzed using chemical or instrumental methods. For example, a liquid/liquid separation of organic components from a water sample and their subsequent concentration to permit trace-level detection by an analytical method might be conducted. Another example is the chemical decomposition of a soil or solid sample using grinding and acidic reduction to free metal species present for analysis.

The sample preparation methods described above are used to produce the actual sample or extract that is analyzed using selective physical or chemical methods. Physical methods include all spectroscopic analysis (e.g., spectrometry, molecular, and atomic spectroscopy) and chemical methods include titrations, other volumetric methods, and precipitation reactions.

The analytical result is used for the determination of the qualitative and/or quantitative composition and concentration of the analyte in the raw sample. After the analysis is complete, critical consideration of the resulting data is carried out, looking for mistakes deriving from preparation and process errors. This is followed by analysis of the raw data for possible interferences. These may be from the method, the nature of the sample or problems from transport. When this is completed the raw data are processed and documented. Finally, the results from all the tests performed are considered together as a whole. Complex matrices and difficult testing problems can all require a combination of methods and analytical procedures to arrive at the final analytical report, so it is important that all the data from a field or project be considered together as a whole at some point.

Considering all data as a whole is also a form of error check; for example, checking for impossible numbers or for sample mismatch of fractions between lab groups. At this point the fundamental question asked in the beginning must again be considered in the light of whether or not the data answer the original question posed by the client. If so, the process is complete; if not, the process must continue until the original question has been answered to everyone's satisfaction.

10.1. DETERMINING THE PROBLEM

Problem determination is a fundamental part of the sampling and testing process and there are multiple aspects to be considered. Since all of the aspects of these two activities can have a dramatic impact on not only the nature of the sampling event but also the sample analysis, it is a very good idea to spend time at the beginning asking simple and basic questions. What does the client want, and what is the question that is being asked? This could be as simple as what the pH and total organic carbon (TOC) level of a groundwater sample is, or as complex as what all the compounds and concentration levels are of all contamination in a soil or water sample.

After the original question is posed it is time to determine the limits of the testing and sampling process by asking yet more questions. The limiting factors for the selection of an analytical method will often be the amount of sample available for sampling and subsequent analysis, and the form or state of the sample to be analyzed. The nature of the sample taken will determine many of the procedures for the sampling event, as well as sample transport, storage, and preparative process and many of the analytical methods and procedures.

As an example of the importance of the sample characteristics the first question asked would be the physical state of the sample to be taken; that is, is it a solid, liquid, or gas? Is the sample homogeneous throughout the field to be sampled? Will the sample be able to be taken in such a way as to maintain homogeneity during transport and sample preparation and analysis? These questions all have distinct and far-reaching effects upon the sampling process and the sample taken (e.g., the process and container used to take and transport an air sample will be much different in form and type from that used to obtain a solid or liquid sample).

The three generally found phases of natural matter on the Earth are gas, liquid, and solid, each having its own sampling techniques and requirements, and each of which will be discussed separately. The fourth phase of matter, plasma, is not usually seen outside an extreme laboratory environment, and while it can be a part of some of the analytical methods [i.e., inductively coupled plasma (ICP) analysis] it is really not a state of matter that can be sampled, much less transported or stored. For these reasons it is outside the scope of this chapter and will not be discussed.

The analytical laboratory's project manger (PM) will ask the above questions of the client. This works as a check to make sure that the test being requested will answer the client's question. The PM maintains direct contact with the client and is responsible for communication between the client and the laboratory. Therefore the PM is the vital link in the initial chain of questions by both the lab and the client. While the PM does not have to be a chemist, he or she will have a broad level of experience and knowledge of the laboratory's methods and processes as well as a detailed knowledge of the specific criteria of the various analytical procedures. This is especially important when it is considered that the PM is acting as a two-way check (e.g., both on the client and his or her requested analysis as well as on the laboratory's response, generated data, and resultant report).

10.2. WHAT IS THE NATURE OF THE SAMPLE?

The nature of the sample or the sample matrix is part of the first general questions asked; however, because of the profound effects this answer has on sampling and subsequent sample preparation and analysis it important to consider it again. Indeed, it is a question that must be asked more than once and be considered throughout the planning of the analytical strategy. The characterization of the sample matrix will often automatically determine many sampling and analytical parameters. For example, the sampling process for a soil is dramatically different from that for groundwater or air. Also, the type and nature of the sample containers are often driven by this question. Precleaned glass containers with Teflon seals and various preservatives are needed for many organic test methods.

Alternately, for metals analysis, plastic or Nalgene-type containers are needed.

Although the three states of matter—gases, liquids, and solids—have their own individual sampling requirements and techniques, there are several key factors in regard to taking a sample that are universal to the sampling process. The first common point is the consistency in sampling. Regardless of the accuracy and precision of the analytical process used, the analytical result can be no better than the quality of the original sample submitted for analysis. Therefore one of the primary concerns of the sampler must be the collection of a true and representative sample. The second point is that once taken, correct and clearly labeled documentation of the samples is essential, as is the accuracy of the data recorded in the field information sheets (Figure 10.3) and in the project notebook.

10.3. THE SAMPLING PROCESS

The sampling event is the first significant result-determining step in the analytical process. The main goal of sampling is to obtain a sample that is representative of the field. The other side of this is to maintain or preserve the sample as a valid example of the field conditions. The possibility of error at that point and the effect of that error are greater than within the analytical methodology [1,2].

One of the most important factors to be considered and maintained in the field sampling process is cleanliness. This factor, along with sample integrity and documentation, is critical to the valid and accurate sampling process. All the sampling equipment used must be thoroughly cleaned (and sterilized if necessary) and stored in such a way as to ensure cleanliness upon initial use in the field. Also, all equipment must be thoroughly cleaned and rinsed between each sampling event if the equipment is reused in the field. It is imperative that sample integrity be maintained both in terms of cleanliness and correctness in composition from site to site. It is very easy to contaminate a whole sampling event with one poor equipment cleaning or by failure to change gloves between sampling a badly contaminated field, a field with low levels of contaminant, or a field that is clean.

Maintaining sample integrity requires that care be taken at all steps of the sampling event to ensure a representative sample is obtained and sent to the laboratory to be analyzed. For water or liquid samples, this may simply require checking for immiscible layers or fractions in the liquid sample. For soil it may require subsampling from various places and thoroughly mixing the subsamples before the final analytical sample is taken. Unless great care

Field Information Form		
		Submission #
Project Name:		Sample Location or ID:
Date & Time:		Sequence No.:
Sample Collector:		Client Name/Address:

Weather Conditions:

Cloud Cover: [] Sunny	Temperature:	Wind speed _____
[] Partly Cloudy	_____	Wind direction __
[] Cloudy	Rain: [] Yes	
	[] No	N
		W ⊕ E
Sample Type:		S

[] Surface Waters: Taken From: [] Shore   [] Boat   [] Bridge   [] Wading

Total Depth _____   [] Surface [] Mid-depth [] Bottom [] Other _____

Type: [] Lake [] Stream [] River   [] Other _____

[] Wastewaters: Start time_____ Finish time _____   [] Composite

                  Sampling Point: _____ Volume _____ ml per [] hr [] ½ hr [] _____

[] Soils /Sediment   Sampling point: _____ Sample Depth _____ [] Composite

[] Drum Waste   Type _____   Layers [] Yes [] No [] Composite [] Grab

[] Other _____ Sampling Point _____ Sampling Depth _____ [] Composite

[] Groundwaters: See well purging information below.

Well Information:	Well Security: [] Lock
Casing Construction Type: [] PVC [] Metal	Reference Elevation _____ft (MSL/NGVD)
Well Diameter: _____ (in.)	Depth to Water _____ ft
Total Well Depth: _____ (ft)	Groundwater Elevation:*__ ft (MSL/NGVD)
*Reference elevation – depth to water = groundwater elevation	

FIGURE 10.3   An example of a field information form.

Purging Information:									
V= (total well depth − depth to water)									
Well Water Volume (V) _____(gal)					Time Purge Began:_____				
Minimum Volume to Purge _____(gal)					Purging Rate (gal/min)_____				
Volume or Water to remove: _____ (gal)					Time Purge Ended _____				
Purging Device:			Lot # _____		Total Volume Purge (gal) _____				
Reading	Time	Cond. (μs)	Temp. (°C)	Turb. (NTU)	D.O. (mg/L)	Sheen/Sal.	Res. Chlor	Color	Ode
1									
2									
3									
4									
5									
Sampling Device Information:									
Sampling Device: _____			Lot # _____		Time Sample Collected _____				
Field Notes: _____ _____ _____									

FIGURE 10.3 Continued.

is taken, however, this method can introduce or compound errors when a comparison is attempted between the subsamples and the field conditions.

10.4. WHAT IS TO BE SAMPLED AND HOW IS IT TO BE SAMPLED?

Once the PM has determined what the client needs to sample and what volume of sample is required for the requested analysis, a bottle kit is prepared and sent to the client or the field for sampling. The bottle kit includes the proper containers as determined by the method used for the requested analysis. (See Figure 8.2 in Chapter 8.) Also included are a set of instructions for sampling and filling out of the chain of custody (COC) forms (Figure 10.4).

Once the sampler receives the sample kit, it is important that it be checked for breakage or content errors before the sampling team goes to the field and begins sampling. Specific instructions will come with the field sampling kit. After the sampling event, the now filled containers are placed in the transport container, often an ice chest. Chain of custody forms, field calibration sheets, field information forms (Figures 10.4, 10.2, and 10.3, respectively), and notes or comments from the sampling event are enclosed in watertight containers (heavy-weight industrial-style Ziploc-type bags work well) and placed in the container. Packing or temperature blanks and enough ice, if cooling is called for, to prevent breakage while in transit is added. Since ice or some other form of cooling is used for most types of samples, it is very important that the lab receive the cooler before the ice melts or the other cooling mechanism is lost. If the sampling takes longer than a day, the samples can be held in the coolers until completion of the sampling event; however, it is vital that the first set of samples taken be maintained in ice during this time.

If samples are held for any length of time, it is important that the hold times as determined by the method are not exceeded (e.g., time from sampling to analysis). Some specific methods as well as many of the wet chemistry methods have hold times that are measured in hours. Most "organic" methods (analysis for organic contaminants) as well as metal analysis methods have allowable sample hold times of days. Table 10.2 gives maximum sampling to analysis elapse times. Once all sampling is complete and the samples are ready to be sent, the contents of all of the coolers must be checked to make sure they are complete. The COC sheets (see Figure 10.4) must also be checked for completeness. This form is then enclosed in its respective individual cooler. The ice level is checked and topped off with ice as needed, and the coolers are sealed. Coolers can often be sent "as is" via a shipper such as United Parcel Service (UPS) or Federal Express, or if necessary can be encased in an outer cardboard shell for shipment to the laboratory.

Usually the only time there are problems with shipping are when not enough ice has been added to the cooler or the cooler has been held up in shipping and is received by the lab after much of the ice has melted, resulting in the temperature of samples exceeding the accepted limits. Another possibility is that not enough ice was added before shipping so the sample containers are not well padded, resulting in container breakage and sample loss.

Chain of Custody Record

Instructions (See back of form)	For Lab Use Only Temp. of Contents: _____ °C Condition of Seals _____	Condition of Contents (or Received on ice, ROI)		For Lab Use Only Submission No.
1. Client (company or individual):	Address: City: State: Zip Code:		Phone: ( ) Fax: ( )	18. Report Type: [] Routine
2. Report to (if different from above):	Address: City: State: Zip Code:		Phone: ( ) Fax: ( )	[] Standard QC [] Data package
3. Client Project Name:	Water Sample Codes (for item 13)	Container Codes (for item 16)	19. Turnaround time [] Standard [] Rush / /
4. Client Project No.:	DW = Drinking	V = VOC	Preservation Codes (for item 15)
5. P.O. No.:	GW = Ground	G = Glass	C = Cool only H = Hydrochloric acid
6. Custody Seal No.:	SW = Surface	P = Plastic	N = Nitric acid M = Monochloroacetic acid
7. Sampled by:	PW = Processed	M = Micro bag/cup	S = Sulfuric acid OH = Sodium Hydroxide
8. Shipping Method:	WW = Waste	O = Other	T = Sodium Thiosulfate

FIGURE 10.4 Sample chain of custody record form.

Item	9. Sample ID or No.	10. Sample Description	11. Date	Time	12. Comp.	Grab	13. Water (Codes)	Air	Soil	Sludge	Other	No. of Containers	15. Preservative / 16. Containers / Analysis Requested	Lab use only Lab Sample No.
1.														
2.														
3.														
4.														
5.														
6.														
7.														
8.														

FIGURE 10.4 Continued.

21. Relinquished By	Date	Time	22. Received By	Date	Time	20. Remarks	For Lab Use Only Sampling Fee _____ hr
1.							
2.							Equipment Rental Fee _____
3.							Profile No.:
4.							Quote No.:

FIGURE 10.4 Continued.

These types of problems must be noted by the lab as soon as the cooler is received and opened. It is vital that any questions, problems, or discrepancies in the coolers and their contents be noted and resolved at the earliest possible time. Often the PM will be able to answer many of the questions as a result of discussions and knowledge of the client's needs without having to contact the client; otherwise, it is necessary to contact the client or field sampler to resolve questions and discrepancies. By doing this at the beginning of the laboratory's log-in process, it is often possible to resolve problems before the sample is released to the various laboratory sections for analysis.

Noting problems early will prevent the sample from being misprepared or analyzed using the wrong methods or otherwise being consumed. It is important to remember that typically field sampling is limited in both time and volume. Often the amount of the sample sent to the laboratory is enough to perform the requested analysis and enough for laboratory quality control needs plus a little more. Therefore there may not be enough sample available to perform multiple preparations and analysis on the sample in question, and thus it is important that the preparation and analysis to be done right the first time.

Other problems in regard to shipping or sample transport could be termed catastrophic loss; for example, the cooler falls from the back of the sampling truck on the highway while at speed, resulting in cooler disintegration, or the tractor at the landfill runs over the cooler, resulting in a very flat cooler. For these types of problems the only answer is to note the loss and resample.

10.5. DOCUMENTATION

As already stated, it is vital that the sample field analysis, project notebook, field calibration, field information, and the COC forms be filled out completely, and in "real time". (See Chapter 8 for considerations of the completion of COC sample labeling and documentation.) It is very difficult to remember all the details in a sampling event while dealing with the sample containers, test equipment, coolers, and all the other associated paraphernalia. To expect to remember all the details correctly and record them later is the height of folly. This folly can have a vast impact on the samples, the data obtained from them, and the final results of the analysis due to field data or sample mislabeling errors. Figure 10.2 is an example of a field calibration sheet for recording analyses such as pH, conductivity, turbidity, and dissolved oxygen. Figures 10.3 and 10.4 are field information sheets and COC record sheets. The field information form (Figure 10.3) and the project notebook are the places in which anything and everything about the field can and should be recorded.

10.6. SAMPLE PREPARATION FOR ANALYSIS

Sample preparation before analysis can range from simple to very complex. A drinking water sample for volatile organic compound analysis might only require the addition of internal and surrogate reference compounds before being introduced into an instrument for analysis. Sample preparation for a more complex sample matrix such as a clayey soil might require grinding, mixing, and subsequent extraction using an organic solvent that produces a concentrated extract for analysis. Preparation might also entail the extraction of the original sample into an organic solvent, this solvent being subsequently extracted with a different solvent that is compatible with the instrument to be used for analysis. Other cases will involve the separation of different fractions of soil or of immiscible liquid fractions and their subsequent separate analysis. The results of the separate analysis are combined on a proportional basis to yield a result representative of the whole original sample. While these are all based on the principles of general chemistry, the specific steps used or performed on the sample are usually specified by a regulatory or government agency that provides the laboratory's certifications and license. (See Ref. 3.)

10.7. ANALYTICAL TESTING

Generally speaking, the natural laws of chemistry and physics serve as the basis for an analytical method. Natural laws are those such as the absorption of a specific wavelength of energy, be it visible, infrared, or ultraviolet, by a specific chemical element or compound. It could also be the ability of a specific chemical compound to be precipitated out of a solution using a specific chemical reaction and subsequently the product being weighed. While the principles used in a method range from the general to the specific, they are all derived from natural laws. This means that an analysis done in one location will give the same reproducible results when done the same way in a different location.

Analytical methods consist of a series of intermediate steps that, when combined, make up the analytical procedure. Therefore, the analytical method could be considered as a set of steps that are determined by the process of measurement, which is in turn based on natural scientific laws. It can include some of the intermediate steps of sample preparation and evaluation and represents the overall concept for the acquisition of data from the sample. The analytical procedure consists of (a) instructions for the initial sampling event, (b) the sample preparation with directions or indications for required equipment and chemicals, and (c) the analytical instrumentation and instrument settings used. The method also gives the required analytical calibration, the area or scope of application, and possible selectivity and/or interferences [4].

On this basis two types of quantitative analysis can be defined: the classic wet chemical methods and instrumental methods. Many of the classic analytical methods still in use today are simple yet reliable "wet" chemistry procedures, including gravimetry (analysis by weight) and volumetric or titrimetry methods. Instrumental methods are based on the chemical nature of the compounds and elements being analyzed for and are separated into organic and inorganic methods. Organic methods are those analytical procedures for compounds that are mostly made of carbon and hydrogen (e.g., pesticides), and generally man-made chemical compounds. Inorganic methods are for non-carbon-containing compounds and such natural elements as the metals.

10.7.1. Gravimetry

If the analytical procedure used to quantify a substance is based on weight, it belongs to the field of gravimetry. This involves precipitating a compound out of the sample or an extract of the sample, drying, and weighing. It is preferred over other methods for microanalysis using sample amounts in the gram range. Its advantages over the instrumental methods to be discussed later include the absence of calibration, thus making it an absolute method. It is a high-precision method and requires limited apparatus. Its disadvantage in a production environment is the large expenditure of time and effort required to perform an analysis. A general procedure for a gravimetric method is as follows: (a) taking of a subsample and weighing, (b) dissolution of the sample, (c) subsequent precipitation of the component of interest according to a prescribed procedure, (d) filtering and washing of the precipitate, (e) drying, and (f) weighing.

A second gravimetric method is called thermal analysis (TA), which is a general term for methods whereby a physical or chemical change in the properties of a substance or mixture is measured as a function of time or temperature. The sample is subjected to a controlled increase in temperature and its change in mass is measured over the course of a preset temperature/time program. A mass change occurs when a volatile component is released from the sample (water vapor, carbon dioxide, organic contaminants, or other similar substances). In addition to decomposition, oxidation can occur, depending on the atmosphere surrounding the sample. Weight measurements are made using a special thermal balance and the resultant data are graphed in terms of temperature versus time and are called thermograms. Each point of change in the thermograph corresponds to a particular reaction of the sample; therefore the thermograph can be unique for a particular sample and thus used as a form of identification.

Anything done to a sample—during sampling in the field, transportation, or storage—that affects potentially volatile components can change the sample, leading to misidentification (e.g., leaving a sample in a sun-heated vehicle could lead to the loss of organic compounds, which could lead to this type of error when a thermal analysis is carried out) [5].

10.7.2. Titrimetry

In titrimetry, compounds or groups of compounds are quantified by measuring the volume of a reagent solution of known concentration—termed the titrimetric solution or titrant—that is required for a defined, complete chemical conversion of the compounds to be determined. Titrant is added until an end point is indicated, and the whole procedure is termed a titration. Chemical conditions required for a titrimetric determination are a defined course of the reaction between the sample and the titrant and the ability to recognize the equivalence point (the titration end point). The type of titration is specified on the basis of the type of reaction occurring [e.g., acidmetry (titrations with acids), alkalimetry (titration with bases), redox titration (transfer of electrons), precipitation titrations (also a form of gravimetry), and complexometric titrations, which are the formation of complexes between the sample and the titrant]. End points are found by a change in an indicator, usually a color change; by measuring pH or changes in pH using a pH meter; or by measuring some other electrical characteristic of the solution being titrated.

10.7.3. Instrumental

As the name implies, instrumental methods require special analytical instrumentation techniques and knowledge beyond the use of balances, burettes, and other glassware. Instrumental analysis consists of a number of steps in two phases: the extraction or sample preparation phase and the analytical or quantitative phase, which is where the sample extract is characterized. The last step is to relate these results back to the original sample and field.

Instrumental methods can be divided into three main groups based on the principle involved. The first group is based on the emission or absorption of electromagnetic radiation (light). Interactions among atoms, molecules, ions, and electromagnetic fields and radiation are used to produce analytical data. These methods are often grouped together as either atomic or molecular spectroscopy, both of which can be either emission or absorption in nature. Atomic spectrometric methods, used for analyzing metals, include atomic absorption spectrometry (AA), atomic emission spectrometry, flame emission photometry, and X-ray fluorescence. Molecular spectroscopic methods, used mostly for organic compounds, include ultraviolet and visible (UV-Vis), infrared (IR), mass (MS), and nuclear magnetic resonance spectrometry (NMR). Mass spectrometry is most often used in combination with a separation method, such as gas chromatography (GC); in this case the method is called gas chromatography/mass spectrometry (GC/MS). This methodology allows the simultaneous separation of a complex sample into its component compounds and the characterization and identification of these isolated individual components.

The second group of methods is based on separation or extraction principles. A sample is prepared and used in the instrumental portion of the method. Many organic methods are based on a chromatographic method for sample speciation and characterization. Chromatography requires the use of a detection method (usually one of the preceding spectrometric methods) for a complete analytical procedure. This means the sample extract is prepared using an extraction method, loaded onto a gas chromatograph for separation, characterization, and speciation of the complex sample components. Once separated, these components are detected and quantitated using a detector.

All extraction methods are based on the physical/chemical separation between different phases, such as partition between liquid/liquid or liquid/solid phases. Separation principles are based on the extraction of a group of compounds from the sample using the principle of differing solubility (e.g., a group of organic compounds will have a greater affinity for a strong nonpolar or less polar solvent that is immiscible with the sample matrix being extracted). An example of this would be a water or soil sample that is extracted using methylene chloride as the extraction solvent. Methylene chloride is not soluble in water, and so organic compounds extracted into it are separated from their original matrix. Table 10.3 summarizes the various types of extraction methods in common use today.

Separation conditions can be varied by changing the pH of the sample being extracted or by using a different solvent. Changing the pH of water changes the protonation $[H^+]$ of the solutes contained in it. For instance, acidification increases the chance that the solute will be protonated, and this means that the solute will have a different net charge. This change in the charge is the reason that the solute compounds have a greater affinity for a solvent of different polarity. This is also an example of the importance of the pH of the original sample during analysis.

TABLE 10.3 Extraction Methods

Name	Abbreviation	Phases	Matrix
Aqueous samples			
Classic		Liquid/liquid, solid/liquid	Liquid or solid
Solid phase	SPE	Absorbent column	Liquid only
Supercritical fluid	SCF	Liquefied gas used in the liquid state under pressure	Solids
Nonaqueous samples			
Solution		Aqueous or organic solvent	Homogeneous liquid
Phase separated		Separate phases and extract using one of the above	Two separate phases

10.7.3.1. Atomic Spectrometric Methods

Atomic absorption spectrometry methods can be of two types—absorption and emission. If a single wavelength of light (monochromatic) passes through a hot gas that contains metal* in an atomic state, some will be absorbed. The specific wavelength of needed light is produced by a hollow cathode lamp (HCL), which has a cathode made of the metal of interest. A sample is passed through an atomizer and into a flame, which produces the hot gas through which the light is passed. The intensity of the analytical wavelength of light is decreased by absorption of the light by the metal in the flame. Any wavelengths not absorbed by the metal are not affected by it. The decrease in the recorded light intensity is proportional to the amount of metal present. Thus, the amount of light absorbed can be related back to the metal present in the sample and from there back to the field.

In atomic emission spectrometry, a wavelength of light characteristic of the metal being analyzed is isolated and recorded. As described above, the sample is passed through an atomizer into a flame where the metal is excited and gives off wavelengths of light specific for that metal. The wavelength of light of interest is isolated using a monochromator and recorded. The amount of light recorded is proportional to the amount of metal in the flame and can be related back to its concentration in the sample and the field.

In both absorption and emission spectroscopy the metal must be excited in a flame. For many elements the hotter the flame the more sensitive

---

* Although this discussion only considers metals, some nonmetals can also be analyzed by this method.

the analysis. There are, however, some similar techniques using the same basic instrument that do not require a flame, although the sample is still heated to high temperature. In all cases a wavelength of light, which may be either in the ultraviolet or visible region of the spectrum, specific to the metal being analyzed for is used in the analysis.

There is another way in which an atom can be excited so that it gives off light of specific wavelengths, and this is by being bombarded by X rays. In this method, called X-ray fluorescence spectrometry, a sample is placed in a beam of high-energy X rays and irradiated. The metals in the sample are excited, absorb energy, and reradiate this absorbed energy as light of specific wavelengths. The wavelength of light given off is specific for specific elements, and the amount of light produced can be related back to the amount of metal present in the sample. In addition to X rays, neutron and gamma radiation can also be used, although the instrumentation is different.

Anything done during sampling, transportation, and storage that changes the metal content of the sample will change the results of this analytical method. This can happen anytime the sample comes in contact with any metal because almost all metals in common use are mixtures. Even if the analysis is not for iron, exposure of the sample to iron could cause it to be contaminated with other metals in the iron. Chromium, nickel, and manganese are commonly found in materials made from iron and could contaminate a sample left in contact with it. Any contact between the sample and any metal should be avoided [6].

The preceding discussion indicates that metals and elements only absorb or emit one wavelength of light. This is not the case. All elements absorb and emit many different wavelengths of light. In some cases, wavelengths from two different elements may overlap or be so close together that they interfere with each other. For this reason it is important for the person doing the analysis to know all the elements likely to be present so that these possible interferences can be controlled or eliminated.

10.7.3.2. Molecular Spectroscopic Methods

Colorimetry includes all the physical–chemical methods that make use of the light absorption of colored solutions for the quantitative analysis of dissolved substances. Colorimetry is based on the direct color comparison without any measuring device beyond the human eye. When light passes through a colored solution its intensity is reduced due to absorption of light of a specific wavelength by the colored substance. This change is used in direct color comparison between the test solution and either vials containing known concentrations of the component of interest or a color chart. The amount of material present is recorded from the reference vial or chart to which it most closely compares.

Two types of colorimetry are common—one in which the solution is colored and one in which a reagent that reacts with the component of interest to form a colored complex is produced. This second method is commonly used in pH determination of soil and water and the levels of chlorine in swimming pool water.

10.7.3.3. Spectrophotometry

Spectrophotometry is similar in principle to colorimetry, except an instrument is used to measure or compare the amount of light absorbed. The instrument consists of a light source and an optical system to isolate the wavelengths of light of interest and pass the light through the sample to a detector. In spectrophotometry light in both the UV and IR regions of the spectrum is used in addition to visible light. This means that components of interest do not have to have perceivable color in order to be detected. In UV spectroscopy, the UV region of the spectrum is used, and absorption of light is caused by the movement of electrons in the molecule. In the IR region of the spectrum, the vibration of bonds in molecules is determined.

10.7.3.4. General Spectroscopy Considerations

From the above discussion it is obvious that any additions, subtractions, or changes in the sample can change the absorption of light and thus change the spectrophotometric results. This is especially the case for colorimetry, in which contamination of the sample with a dye or other colored compound can have a dramatic effect on the quantitative results. Also in such analyses, if the cells used to hold the sample in the instrument are dirty, they can interfere with the analysis. Likewise, a cloudy sample will give an erroneous result.

All analytical methods that depend on the absorption or emission of light have several requirements in common. They all need blank solutions that contain all the components in the test solution except the component of interest. These blanks are used to zero the instrument before readings are made. The second requirement is a standard or calibration curve. These are prepared by making a series of solutions containing different known amounts of the component of interest. These are then measured and a graph of the amount of component present versus the instrument response is prepared. Using this graph the concentration of analyte in an unknown can be determined. Statistically this standard curve should have a regression coefficient r^2 of 0.99 or 99%. This shows that the graph is nearly a straight line and that there is a good relationship between the actual amount of material present and the instrument reading [7].

10.7.3.5. Mass Spectroscopy

Mass spectroscopy is different from other spectroscopy techniques in that it measures the mass of molecular ions or fragments produced by ionizing the sample. The fragments produced from a compound are unique, and can thus be used for its identification. A sample of a compound is introduced into the mass spectrometer, and the ions formed are separated in the analyzer using either a magnetic or an electrical field. The separated ions and fragments are recorded and their unique fragmentation pattern is used for identification. When this same fragmentation pattern is obtained from an unknown in a mass spectrum, it shows that this compound is present. Because mass spectroscopy is best used to determine the composition of a pure single molecule, it is often used after chromatographic separation of a mixture into it component parts [8,9].

10.7.4. Separation Methods

Initial physical and chemical separation may be necessary as adjuncts to final separation and determination methods. Two reasons for separation are:

1. If the selectivity of the determination method is insufficient and interfering impurities must be separated out and the sample concentrated before analysis.
2. If as many substances as possible are to be qualified and quantified in a single procedure. In this case, the sample must be separated into its respective components (e.g., aqueous/nonaqueous, solid/liquid).

Separation methods can be initially divided into two groups, depending on whether or not they act on the basis of analyte transformation. Transformation includes the classic gravimetry, electrogravimetry, which is when a substance is transformed with the aid of an electric current, and volatization of a substance in an aqueous phase.

Separation without substance transformation can be accomplished by partitioning on the basis of partial charges, masses, sizes, or shapes, and different vapor pressures of the compounds of interest. Separations caused by partitioning between two immiscible phases are also very important. Depending on the types of components and solvents involved one differentiates adsorption or desorption processes (solid–liquid or solid–gas), liquid–liquid partitioning (usually as liquid extraction), ion exchange (between a solid ion exchange matrix with a substitutable charge group and a liquid phase), and liquid–solid extractions.

Specific methods of separation based upon particle charge depend on the differences in the migration of free charged particles while in an electrical field. Differences in the particle mass, size, and shape also make it possible to separate substances by such methods as gel filtration or centrifugation. Separations can also be based on vapor pressure differences between compounds using distillation, condensation, sublimation, and crystallization.

For extraction of solids, extraction desorption concepts apply and may be of several different types. An aqueous solution may be used to extract inorganic components and ions from a solid material for subsequent analysis, or an organic solvent may be added to a soil or solid matrix either before or after the sample has been dried with a chemical agent such as anhydrous sodium sulfate. To ensure good chemical mixing and extraction the sample/solvent combination is mixed for a period of time either on a mixer or using an ultrasonic device. Ultrasonic mixing is performed using either an ultrasonic bath or horn. During this procedure solvent aliquots are alternately added to and decanted from the sample after each period of sonication, and placed into a single vessel for future concentration.

Aqueous extracts are often used as is, while organic extracts are often concentrated. Organic sample extracts are concentrated using a "blow down" device where a regulated stream of dry, pure gas is passed over the vessel top in such a way as to promote evaporation of the solvent without disturbing the liquid surface. The gas blow down method of concentration is better than the older methods of "cooking off" the excess solvent using special glassware and a distillation column on a warm water bath. Boiling methods are time- and labor-intensive, and even when used with great care, are prone to catastrophic loss through "bumping" or explosive boiling and leaking or broken glassware. After the sample extract has been concentrated to a set volume (usually 1 ml), it is stored in a small vial sealed with a Teflon-faced septum or cap at 4 °C until instrument analysis takes place.

Another general form of extraction is what is termed gas/liquid or gas/solid extraction. These are both variations upon the principle of headspace analysis. Headspace analysis is the analysis of a gas that has been in contact with a liquid or solid sample and is used to draw conclusions about the nature and/or composition of the underlying sample. There are two general variations of headspace extraction; the first and most frequently used form is where an inert gas, such as helium, is bubbled through a liquid sample, thereby stripping volatile analytes from it. This is often referred to as purge and trap (P&T) or dynamic headspace analysis. The stripped gas sample is passed through an absorbent trap, which retains the volatile analytes. The collected analytes are analyzed by rapidly releasing them from the absorbent trap (usually by heating and back flushing) and introducing them as a "band or slug" into a gas chromatograph.

Gas chromatography, discussed below, is particularly well suited for this analytical method because the extracted sample is already in a vapor or gas phase and can be directly introduced into the GC's carrier gas stream for the separation.

The other form of headspace extraction is where the sample—be it liquid or solid—is placed in a sealed container with a gas volume above it. The container is then heated to a set temperature and allowed to come to equilibrium between the two phases (i.e., solid and gas). After a prescribed period of time a fixed volume of the container's gas phase is removed and introduced into the GC's carrier gas stream and analyzed [10].

The last general form of extraction is known as solid phase extraction (SPE). The use of this form of extraction is rapidly increasing because of the smaller volumes of solvent used in the extraction and the concurrent ease of sample concentration due to the smaller original volumes involved. It allows for high selectivity in the components extracted and retained and on the elution solvents, which are used to remove the component of interest. A liquid sample is passed through a short disposable column containing an absorbent material using either pressure to push or vacuum to pull the liquid through the column (Figure 10.5). By choosing suitable column material and solvent, the selective retention and subsequent elution of analytes of interest can be obtained, providing both sample cleanup and concentration in one or two steps. This extraction method is only usable for liquid samples (usually water) that have little or no particulate matter because suspended solids will clog the extraction column before the entire sample has passed through.

Any material accidentally added to a sample during sampling, transportation, or storage, or any error in sampling would also be "concentrated" during the above procedures. Even small, seemingly inconsequential contamination can be magnified 10 or 100 times during

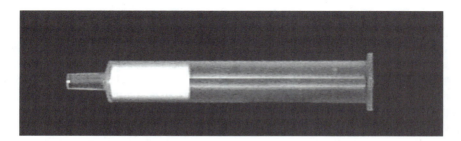

FIGURE 10.5   An extraction column used for purification of analyte before analysis.

concentration steps. When this happens it can have a pronounced detrimental effect on the analytical results and may produce entirely erroneous results, resulting in dramatically increased costs of analysis and remediation [11–13].

10.7.5. Chromatographic Separation

There are a myriad of chromatographic methods in common use, all of which are extremely powerful separation techniques and all of which operate on the same basic principles. They have a stationary phase over which a mobile phase moves. When a complex mixture of components is introduced into a chromatographic procedure, the components are differentially partitioned between the two phases. Differential partitioning results in separation of the components as they move under the influence of the mobile phase. Once separated the individual pure components are identified and quantified.

Two specific types, liquid chromatography (LC) and GC, are most commonly used in environmental analysis. Liquid chromatography, also known as column chromatography, is performed in columns using only the pull of gravity to move the mobile phase. The columns used have an inner diameter of 1cm or greater and a stationary phase with particles between 100 to 300 µm in diameter. The second kind of LC uses high pressure to move the mobile phase through the column. Because of this it is called high pressure or high performance LC (HPLC). High performance LC uses narrow columns 2 to 4 mm in inner diameter with particles of 3 to 10 µm in diameter as the stationary phase and mobile phase pressures up to 4 MPa. High performance LC achieves considerably higher separation efficiencies than does column chromatography.

As with the SPE method discussed above, particle-free samples are required for either of these methods. Suspended solids will be "filtered" out on the top of the column and prevent the movement of mobile phase through it.

The other chromatographic method commonly used in environmental analysis is GC. In this method, a carrier gas such as helium, nitrogen, or hydrogen is used as the mobile phase to drive the sample through a column that contains the stationary liquid phase. The sample—which must be in either a gaseous or a vapor state, vaporization being accomplished by heating the sample when it is injected into the GC—is then separated by partitioning between a liquid phase and the carrier gas. A limitation of this method is that the sample must be able to be converted to a gas or vapor state with minimal loss or degradation.

The sensitivity of modern gas chromatographic methods cannot be overemphasized. Solvents being used in a part of a laboratory several rooms—or in some cases several floors—away may be detected in a sample being analyzed. Thus great care must be taken to ensure that samples are not exposed to any vapors during sampling, transportation, and storage. This is also why both positive and negative controls are needed for all methods.

In both HPLC and GC it is possible to "identify" a compound by its retention time (e.g., the time between when a compound is introduced into a column and when it emerges). If repetitive analysis of samples that all have the same constituents and the instrument is set up the same way retention times can be a very good indication of the type and amount of compound present. However, two compounds can have the same retention times, and so an independent separate identification method is always appropriate. Gas chromatography/MS is extremely valuable to the analyst and laboratory because the compounds can be identified conclusively during the analysis and it does not require costly and time-consuming conformation using a second method.

Introduction of samples that will not vaporize or decompose when heated will destroy the chromatographic column. Any material in the sample having these characteristics will invalidate the analytical results, prevent further analysis, and prevent the use of the instrument for several days. Also, gas chromatographic columns are not cheap, sometimes costing over $1000, and thus care needs to be exercised in the preparation and introduction of samples into a GC to ensure its suitability and continued operation [14,15].

10.7.6. Combined Methods

In modern instrumental analysis it is common to use a combination of instrumentation in tandem or sequentially. The most common is GC coupled with MS (GC/MS). As separated compounds exit a gas chromatographic column they are introduced into a mass spectrometer. This method thus not only separates a complex mixture into its component parts but also identifies the components, making it an extremely powerful analytical tool.

The output from a GC/MS is sent to a computer, where it is recorded and displayed. The computer also contains a library of the unique fragments of a wide range of compounds. It can be used to search this library for compounds with the same fragmentation patterns as the unknown component from the chromatographic column and thus identify it. In one analytical procedure taking little time, a very complex mixture can be separated and its component parts identified.

There are two very important limitations to this method that must be kept in mind at all times. It is possible that two different compounds will come out of a chromatographic column at the same time. It is also possible that these two compounds will have similar fragmentation patterns. When this happens the computer output can misidentify the compound exiting the chromatographic column. This is a particular problem when the GC and MS are set up for a particular analysis and, without changing the instrument parameters, it is used for a very different analysis [9,16].

10.7.7. Electroanalytical Analysis

Electroanalytical methods use an electric current (i.e., amperage and voltage) or a change in potential to produce an analytical result. These methods are often based on the change in a material that occurs upon the application of an electrical potential. They are commonly used as the indication method in such classic wet chemistry methods as titration processes. Electroanalytical methods use ion and electron movement and phase interfaces of electrodes and ions to obtain information about the type and concentration of organic or inorganic material present in a solution.

The two main electroanalytical forms are transformation of electrical energy into chemical energy (i.e., electrolysis with current flow across a cell) and the transformation of chemical energy into electrical energy using a galvanic cell. As an example, electrolysis takes place when current is applied to a copper rod immersed in a copper solution and is connected with a second half cell made up of a platinum rod immersed in an acidic solution. Copper ions will take up electrons and be deposited as metal on the cathode. A galvanic cell (wet cell) is made when of two half cells, each consisting of a metal electrode immersed in its cationic solution, are connected to each other. The measured potential of this cell is the difference between the individual potentials of the electrodes in their solutions. If one cell is copper ions in a copper sulfate solution and the other a zinc rod in a zinc sulfate solution, the connection between the solutions and rods makes a galvanic cell. The reduction of copper ions to metallic copper at the copper rod and the oxidation of the zinc metal to zinc ions at the zinc rod takes place. This combined reaction generates a current.

Testing methods employing the principles of electroanalytical chemistry are mainly limited to the two forms known as potentiometry and conductometry.

In potentiometry, the characterization of the amounts and types of the materials present are determined by measuring the potential differences between a measuring or indicator electrode and a reference electrode using a constant potential. This is the principle used to measure the pH of a sample.

The hydrogen ion concentration of a sample is measured using a pH electrode in combination with a reference electrode or solution. The pH electrode usually consists of a glass tube that contains a buffer solution of a known pH and a reference electrode. In operation, pH measurements are made using a single-tube measuring device, containing both a pH and reference electrode, which is immersed in the sample to be tested. The glass wall of the electrode at the end of the tube acts as a complex membrane and allows hydrogen ions from the sample to penetrate into the glass electrode. Sodium ions are displaced by hydrogen ions and a potential difference develops at the glass/water interface. The exchange rate between the sodium and hydrogen ions on the surface of the glass of the electrode is pH-dependent. Therefore the different potentials between the reference solution and the sample are linked to different pH values, and it is this difference that is measured and interpreted as the pH of the sample [4].

In conductometry, the presence of ions in solution allows a current to flow when an alternating current voltage is applied. The conductance or resistance of the cell is the sum of the ions across the sides of a reference cell in comparison to the sample cell. Measurements using conductometry are mostly used in analytical methods in which a large change in conductance of a sample cell is compared to a reference conductance. Conductance measurements are used to measure the salt content of aqueous solutions or extracts.

Any error in sampling that changes the pH or salt content of the sample will adversely affect these measurements. It will also affect the proposed cleanup and remediation procedures, and it may interfere with other analytical procedures [17].

10.8. ANALYTICAL METHODS—SOME SPECIFICS

In determining the actual methods (described above) to be used in the sample analysis the first thing to consider is the nature of the sample; that is, is it solid or liquid and what is the client's question? This consideration is a way to start the logic tree for the analysis, which is a relatively simple, logical consideration of the sample and what you want to learn from it. Historically it has been repeatedly found that a few minutes' time given to such a basic consideration of common sense before the sample is taken (much less logged in for analysis) can save time, effort, and frustration later.

Ideally this consideration should be second nature to the PM in discussing the problem or analysis needed with the client. (Remember that the PM is the first and last contact with the client and hopefully can get the complete picture.) If a client thinks his well is contaminated with organics from the next-door neighbor or some governmental or regulatory agency has requested monitoring of an underground storage tank (UST) the base method to be used for analysis of the sample is determined by this request. If the sample is being submitted for initial characterization (i.e., what it is or what is in it), basic analytical methods for the general component determination are used. The use of GC/MS methods in both these cases is highly recommended because of the ability of this method to isolate and determine many different analytes with one analysis.

10.8.1. Methods for Liquid Samples

Water or liquid samples are measured by volume, the pH is adjusted, and an extraction reference is added. (This extraction reference is traditionally called a surrogate, and is a known compound or group of compounds with similar characteristics as the target compounds.) This way there is a measure of extraction efficiency for the sample; that is, if X amount of a surrogate with the same or similar characteristics to your target analyte is added to the sample and after extraction the analysis of the sample extract shows a 90% extraction efficiency it could be assumed that there was a 90% recovery of the target compound. Unfortunately this does not hold to true for many compounds. Usually extraction efficiency should *not* be used to correct the analytical results for extraction efficiency; rather, the surrogate recoveries are used in comparison to the method-determined limits to validate it. In the case of an invalid testing event the sample may require reanalysis to confirm matrix or method interference. If enough of the original sample remains, the entire sample preparation and analysis process may be repeated.

10.8.1.1. Extracting Liquid Samples

Liquid samples are extracted in one of the following ways, depending on whether they are aqueous or nonaqueous. (Extraction methods are summarized in Table 10.3.)

Aqueous Samples:

Classic—Liquid/liquid or solid/liquid (i.e., liquid sample matrix/ organic solvent or solid matrix/organic solvent).

Solid phase—A liquid sample is passed through an absorbent that retains the analyte, which is flushed from the column for analysis.

Supercritical fluid—A sample is extracted by the use of a gas that has been compressed to its liquid phase. This is usually carried out using liquid carbon dioxide (CO_2) [10].

Nonaqueous Liquid Samples:

Small quantities of oil or some other nonwater miscible matrix or a
solid—extracted with either water or an organic solvent.

Large quantities of oils or nonwater immiscible matrixes are usually
separated from any water phase either by decanting or by using a
drying agent and then extracting the remaining material, usually
with either water or an organic solvent.

10.8.1.2. Multiple Extractions

Multiple sample extractions can be carried out in two ways. The first is the
extraction of the sample with multiple aliquots of solvent, usually three
times. For example, three 50-ml aliquots of methylene chloride are added to
1 liter of water sample, each aliquot is shaken for 2 min, is allowed to
separate, and is collected combined with other extracts for concentration in
a common container. Theoretically 50–75% of the sample contaminant will
be extracted by the first aliquot. Lighter or less complex sample matrixes
might have an extraction efficiency of upwards of 90% in the first fraction,
while heavy molecular weight compounds or complex matrixes or soils may
have a far lower efficiency. This will be indicated by recovery of the
extraction surrogate. The standardized use of three solvent aliquots with a
fixed solvent contact time of 1 or 2 min acts as a point of standardization for
the extraction process, hopefully ensuring that if you have x amount of a
compound in your sample you will have extracted it all. If 100% is not
extracted because of matrix interference you will be aware of the decreased
extraction efficiency by the monitoring surrogate extraction.

10.8.2. Methods for Extracting Solid Samples

Solid samples are extracted by adding an extractant, then mixing and either
decanting the extractant or filtering to separate solids from the extractant.

10.8.2.1. Methods for Solid Samples

Individual samples are taken from the field using a predetermined pattern
and extracted. The extracts are then analyzed. The handling of these solid
samples will be illustrated using soil as an example.

A soil sample is characterized and mixed to assure a uniform sample
from the sample container. A subsample is weighed and the original weight
being extracted is recorded in the laboratory logbook. A portion of the
sample must be analyzed for its percentage of solids because this factor must

be applied to the resultant data for the soil sample to yield the field's true component concentration. Extraction surrogates are added and the sample is mixed to ensure even distribution. Mixing with a drying agent, such as sodium sulfate, dries the sample.

An aliquot of the extraction solution is added and the sample is mixed for a fixed time, either by stirring or through the use of an ultrasonic extraction device. Solvent is poured off through a piece of filter paper into a flask and the sample is re-extracted twice more for a total of three solvent aliquots. The extract can be passed through a cleanup phase; that is, a fluorasile cleanup column (see Figure 10.5) that has been prerinsed with the solvent used for the extraction. This or the original solvent extract is reduced in volume to 1 or 2 ml and analyzed.

10.9. DATA ASSESSMENT

The data produced using one of the analytical procedures described above is what is termed "raw" data and must be related to some standard or reference to determine the relationship of the analytical response to an actual value. This is usually done by comparing the raw response to the response obtained from the instrument or method when a series of known standards is analyzed. The instrument responses from these standards are used to prepare a response or calibration curve for the instrument that encompasses the linear response range of the method. Often the range and number of these calibration standards as well as the statistical form used to generate the response curve will be dictated by published method conditions.

Ideally, the raw sample response will fall within the range of responses used in determination of the instrument or method calibration range. If the raw response is outside this range, there are two factors to consider. The first is what the concentration limit is for the particular component or compound in the analytical method used. If the analyte in question exceeds the calibration range of the analytical method or instrument, the answer may be to dilute the sample by an accepted factor and reanalyze.

Diluting and rerunning a sample can introduce contamination from the diluting agent, and it will introduce a dilution factor that must be added to the calculation when deriving the true concentration of the sample. Dilution may mask or reduce other important analytes in the sample to such an extent that they are not accurately determined using the analytical method. Thus, great care and high-accuracy glassware and measurement must be used in making dilutions. A general rule related to all analysis is that any additional manipulation of the sample will introduce error.

Usually there are also other points to consider as part of this. One of these points is the question of how far above the instrument limit the analysis went. This is important because the instrument may now be contaminated. Contamination can be determined by running a method blank to see if it is below detection levels for the compound in question. If it is, there is no problem. However, if it is above detection limits, the analytical system requires cleanup or maintenance to bring the level back below the detection threshold for the compound in question. Finally, the system must be thoroughly checked using a series of system blanks to ensure conformity to method requirements. In extreme cases the system might require further cleaning or component replacement before the instrument system is clean.

Another question to be considered is if the level of the analyte interferes with other compounds that might be present. If so, the other compounds may not be reliably seen or reportable. The sample in this case may need either dilution or further separation and reanalysis with the reporting levels modified to reflect the dilution factor.

In extreme cases, the level of a compound may exceed its solubility in a sample or solution. When saturation occurs excess material will separate out, as in the case of oil or gasoline contamination of groundwater. Compounds that make up the gasoline dissolve in water until saturation occurs, and the remainder of the gasoline then pools or forms a layer on the top of the water sample. This can cause errors in the characterization of a sample if part of the nonuniform sample is taken to represent the entire sample. This is an example often seen in groundwater samples from a field contaminated by a leaking gasoline storage tank. While the layer of free product on the surface may lead to very high concentrations for compounds such as benzene or toluene in the analysis of the water sample, they are often not the true concentrations of the compounds actually dissolved in the water.

This is an example of a sample with multiple layers in which the concentration will vary between the layers of the liquid sample. In the case of a soil, the analytes may, in extreme cases, be seen as deposits within the soil matrix as a whole. Because this is not a characteristic sample, it can lead to analytical problems unless care is taken in the original sampling of the site and the subsequent sample preparation, extraction, and analysis.

In some cases the level of contamination may be below the maximum contamination level (MCL), and in this case further action beyond continued monitoring and assessment is usually not required. If the component is below detectable levels it is probably also below MCL. If not, the sample extract may need to be concentrated to obtain analyte for analysis. As with diluting, as discussed above extreme care must be exercised in the concentration of samples.

10.10.   REPORT GENERATION AND DOCUMENTATION

What is the overall report produced by the laboratory as a whole? It can be as simple as a handwritten or typed transcription of the verbally reported results discussed with the client by the PM. However, most analytical testing labs today use some form of computer database to produce the reports. Generally these are known as a laboratory information management system (LIMS), and act as an aid both in handling the large volume of samples received and the data generated and in collating and generating reports. One soil sample for a general screening for organic and inorganic components may have several hundred individual results linked to it. Because the chance for transcription errors increases every time the data are copied, it is better from a data integrity standpoint to enter the data into the LIMS either directly from the instruments or from the original data sheets. The data are then double-checked, peer reviewed, and subsequently checked by a supervisor or a group leader to ensure accuracy. The importance of double-checking the results and the data entered into the LIMS system cannot be stressed enough.

Once all the requested analysis is done and all the data have been entered into the LIMS system, the PM should review it for gross errors and impossible numbers. After this is done and any problems are resolved, a draft report is generated and reviewed. If all is acceptable, a final draft is printed and again checked for errors or inconsistencies by the quality assurance/quality control (QA/QC) offices and finally by the lab manager or director who is the final check on and ultimately responsible for the data and the report produced by the lab. If everything is acceptable, the report is signed by the final review authority and is released for delivery to the client.

Often in today's work environment the report is sent as a faxed copy or an electronic data deliverable form (EDD) via the Internet or data media. However, it is still important that a hard copy be sent to the client unless specifically directed otherwise. The reason for this is that the final hard copy is the legally accepted copy of the data if there are any complications or problems. The laboratory results can have far-reaching and often very expensive effects. For example, the ruling of a drinking water system out of compliance because of contamination could result in legal action and fines. The classification of a field as contaminated can result in the current and possibly prior landowners being liable both legally and financially for the contamination and the subsequent cleanup costs.

Because of the potential ramifications of the lab's results, it is imperative that the lab maintains internal records and documentation of the methods and results and the actual raw data, both in electronic and hard form, as well as any problems or variations from the accepted methods and procedures that might have occurred. Many regulatory agencies require that the data associated with a drinking water system be available in time frames of multiple decades. Also, it may take months for the client to get all of the data (results) from field sampling and to collate and generate its own report. This means that the actual data must be available for a time frame of at least a year [18].

10.11.   DO THE RESULTS ANSWER THE CLIENT'S QUESTION?

While this question may seem redundant, as stated earlier, it is important to consider it often during the testing and report-generation process. This is due to the fact that the answer to the client's original question will often lead to many more questions and problems that will need to be considered and answered in turn. Thus it is important to come back to the original question and be certain that it is answered.

10.12.   SOME GENERAL QUESTIONS AND CONSIDERATIONS

What if the concentration is above the MCLs? Traditionally this means that there is a contamination problem that must be resolved. Usually a government or regulating agency at the local, state, or federal level will become involved. For instance, in the United States a department of the Environmental Protection Agency (EPA) will have already acted to set levels of acceptable contamination. These levels will vary with the compound in question according to the nature and potential danger of the contaminant, and are often termed the maximum levels of contamination. Another term often used is that of action level; that is, that concentration level above which some form of remediation or site cleanup is mandated. If the level of a contaminant exceeds the MCL level, the field will often require mandatory cleanup according to the various state or federal regulations. Once the federal or state-level EPA becomes involved and the levels are confirmed, the source of contamination is determined, along with the extent of contamination and its possible hazards. Fines are levied via the courts—hopefully, to pay for the required site cleanup and to prevent future spillage or accident by encouraging better management, technology, or systems.

States and countries other than the United States can and do set MCL levels that are different from those of the United States Environmental Protection Agency (USEPA). It is always essential to determine the locally applicable rules and regulations.

10.13. CONCLUSIONS

The first question to ask before analysis of a sample can occur is what it is that the client wishes to learn and what the nature of the sample is. There are both simple and complex analytical procedures that can be applied to samples in order to answer the client's questions. Some are methods carried out in the field and some are done in the commercial laboratory. Field analysis such as pH and dissolved oxygen are simple and quick and require simple instruments. More complex analysis involving extraction, concentration, and instrumental analysis are done in the commercial laboratory. In any analytical procedure sample preparation is an essential first step that is often followed by extraction and finally by analysis. Analytical procedures may be standard wet chemical methods or more complex instrumental methods. Commonly the instruments used fall into two groups—those based on spectroscopy and those based on chromatography. In some cases a combination of spectrophotometric and chromatographic methods will be used in the analysis. Once all the analysis is completed the data are checked and a final report is prepared and presented to the client.

The final data and resultant report can be no better than the original simple submitted for analysis, and this also applies to all steps of the analytical procedure. If care is not taken at every step—sampling, transport, storage, and during the analytical processes—errors will be compounded and the resultant data will be flawed. The ultimate result is that the data do not accurately reflect the true nature of the field being tested.

QUESTIONS

1. What are the limiting factors for the determination of sampling and testing methods?
2. Insulated ice chests are often used to transport samples from the field to the laboratory. What must the ice chest or cooler contain in addition to the samples? What two purposes does the ice serve?
3. What is one of the most important factors in obtaining a representative sample in the field? What two factors go along with this point?
4. Name three common methods used for sample preparation.
5. What laws are fundamental as applied to an analytical procedure?
6. What are the two types of quantitative analyses and three groups of instrumental techniques? Name each and give a short description of how they work.
7. What does PM stand for? What is this person's role?
8. When the concentration of an analyte of interest is far beyond the range of the instrument or method, what must be done to analyze it?

How can analyzing samples containing too much of an analyte adversely affect the subsequent analytical procedures?
9. What is the sampling logic tree and how is it used in sampling and sample analysis?
10. What is an extraction reference or surrogate and how is it used in the sample analysis process?
11. What are three common methods of sample extraction?

REFERENCES

1. Description and Sampling of Contaminated Soils: A Field Pocket Guide. EPA 625-12-91-002. Cincinnati, OH: National Service Center for Environmental Publications. Fed Reg 1991.
2. Corbitt RA, ed. Standard Handbook of Environmental Engineering. New York: McGraw-Hill, 1988.
3. Sampling: When, Why, How. http://www.epa.gov/region/offices/oea/ieu/manual/inspect08.htm.
4. Brady JE, Humiston GE. General Chemistry: Principles and Structures. New York: Wiley, 1982.
5. Haines P, ed. Principles of Thermal Analysis and Calorimetry. London: Royal Society of Chemistry, 2002.
6. Dean JR, Ando DJ. Atomic Absorption and Plasma Spectroscopy. New York: Wiley, 1997.
7. Pavia DL, Lampman GM, Kriz GS. Introduction to Spectroscopy. Ann Arbor, MI: Brooks Cole, 2000.
8. De Hoffmann E, Stroobant V. Mass Spectrometry: Principles and Applications. New York: Wiley, 2001.
9. McMaster MC, McMaster C. GC/MS: A Practical User's Guide. New York: Wiley, 1998.
10. Kolb B, Ettre LS. Static Headspace–Gas Chromatography: Theory and Practice. New York: Wiley-VCH, 1997.
11. Dean JR. Extraction Methods for Environmental Analysis. New York: Wiley, 1999.
12. Taylor L. Supercritical Fluid Extraction. New York; Wiley-Interscience, 1996.
13. Simpson NJK, ed. Solid-Phase Extraction: Principles, Techniques, and Applications. New York: Marcel Dekker, 2000.
14. Poole CF. The Essence of Chromatography. New York: Elsevier Health Science, 2002.
15. Kitson FG, Larsan BS, McEwen CN. Gas Chromatography and Mass Spectrometry. New York: Academic, 1996.
16. Bard AJ, ed. Electroanalytical Chemistry. New York: Marcel Dekker, 1999.
17. Nakagawa AS. Lims: Implementation and Management. London: Springer Verlag, 1994.

11

Traps, Mistakes, and Errors

Alfred R. Conklin, Jr., and Rolf Meinholtz

Traps, mistakes, and errors can occur at any stage in sampling and analysis; for example, in the field, during transport and storage, during analysis, and in the interpretation and communication of analytical results. Traps occur when incomplete data about a sampling activity, including the history and characteristics of a field, have been collected or something unusual happens. Mistakes often are the result of untrained or inexperienced personnel who do not understand the procedure, the importance of each step, how it works, and what results are reasonable and expected. Mistakes could also be made by personnel who simply do not pay attention to what they are doing.

Most traps, mistakes, and errors associated with environmental sampling occur with the physical sampling process as it occurs in the field. Fewer occur during transport and analysis, and the least with chemical and statistical analysis and reporting of data. In some cases errors are caused by the heterogeneous nature of the environment. In most cases they are caused by personnel [1]. Unfortunately nothing will completely eliminate errors and mistakes, but they can be minimized by having checks at each stage of the process, by having good communication between samplers and laboratory personnel, and by using statistical tools. To be effective these must be carried out with careful attention to detail.

The first place in which personnel have trouble is in reading maps and judging distances. A way around these problems is the extensive use of a global positioning system (GPS). The second place in which problems occur is in transporting and storing samples. The thoughtful use of chain of custody forms and procedures will minimize problems associated with these activities. The third place will be in carrying out the analysis, and the last in reporting the results. Extensive and continuing training in analytical procedures and reporting protocols will minimize these problems. Good communication between personnel involved at each stage is also crucial in minimizing errors.

During transportation and storage, problems arise due to mixing samples and putting the wrong paperwork in the sample transport box, along with breakage and the loss of temperature control. Carefully choosing sample containers and packing materials, paying attention to details in completing paperwork, including labeling, and including the correct paperwork with the proper samples will all go a long way toward minimizing problems. Checks and double checks at this point can prevent the loss of samples and incorrect laboratory analysis.

In the analytical process there are also a number of traps and errors that can be made. Samples may be extracted using the wrong extractant, or the correct extractant might be used but with the wrong instrument. Interfering compounds or species present in the original samples or added later may result in inaccurate results or hide the compounds of interest altogether. The laboratory technician may incorrectly read or record data. With sufficient numbers of controls—both positive and negative—these types of problems can be caught and rectified.

In searching for errors often the first approach is to apply statistics rigorously to the field samples and their analytical results. While statistical analysis is critical in any sampling program, it is not sufficient in and of itself to ferret out all problems. An example of statistics not being applicable to a specific set of data is given in Section 11.2. A better first approach is to look for possible sources of errors recorded in the project notebook and in the chain of custody, field calibration, and field information forms. A summary of errors and mistakes made both in sampling and in analyzing samples is given in Table 11.1.

11.1. HISTORY OF THE AREA

Because it is one of the first pieces of information obtained about a field, its history is a beginning source of mistakes. Failing to obtain a complete history of an area can result in costly and time-consuming errors in interpreting data and developing remediation plans. A complete and thorough investigation of all aspects of an area will be invaluable in

TABLE 11.1 Sources of Mistakes and Errors Made in Sampling and Analyzing Environmental Samples

Activity	Sources of error
Field sampling	Incorrect field, incorrect position in field, inconsistent sample size, failure to record in field notebook (also see Table 11.3), incorrect labeling, failure to fill out proper forms, improper or failure to preserve (e.g. acidification, cooling)
Transportation	Incorrect sample container, loss of sample, contamination from other samples or surroundings, incorrect labeling, incorrect shipping container and container conditions
Storage	Incorrect storage conditions (e.g. temperature, isolation, sunlight, time)
Analysis	Incorrect storage conditions, incorrect handling (e.g. wetting, drying, heating and cooling, exposure to contamination, loss of contaminant), incorrect sample preparation, wrong extraction method, incorrect solvent used
Reporting	Incorrect input of data, incorrect mathematical and statistical analysis

explaining both expected and unexpected analytical results. Examples of the importance of obtaining the history of the field to be sampled have been given in previous chapters. (See particularly Section 3.6 in Chapter 3.)

A field history can be of great help in determining what potential contaminants may be present. For example, if the groundwater is being tested at the site of an old gasoline station, it is within reason that gasoline contamination may be found. Also, and just as important as the initial field characterization, is the ongoing history of a site. This can be as fundamental as a field map indicating where the sample points are located to as complex as the field sample history plotted as a thread over time with reference to field soil characteristics and hydrology.

The biggest mistake made in obtaining the history of an area is the assumption that it was not or could not have been used for a specific or general activity that would lead to contamination. These assumptions are often made by looking at the area as it is presently and seeing no indication of human activity. Of course this view leaves out the possibility that somewhere below the surface is information about previous uses, which may be indicated by the occurrence of some common soil components. The occurrence of higher than normal levels of elemental carbon may indicate the existence of campfires or other burning in the area. This might be caused by normal occurrences, but may also indicate human habitation. Higher than expected levels of phosphorous in soil may indicate the same thing.

Thus, exploration of the history of an area must be done with an open mind, and it must never be assumed that the area is not and never has been contaminated. The most unlikely fields, often remote from populated areas, are contaminated in one way or another. A small area enclosed in trees in a remote field on a farm was found to be an area in which a farmer had dumped both empty and partially filled containers of excess agricultural chemicals, including toxic compounds, some dilute and some concentrated. Such an area is a source of contamination to surrounding land and water supplies. Another remote field surrounded by trees was found to contain a pit being used as a dump. The area was both remote from a population center and screened from surrounding roads by the trees. Gullies are commonly used as places to dump unwanted materials. Often this is done for two purposes. One is a mistaken idea that this activity will decrease erosion and slow or stop the development of the gully. The other reason is that the material is not visible to the casual passerby. Even if the discarded material is buried it will be exposed later when further erosion has occurred.

In terms of human occupation, some chemicals are commonly found, that were used in earlier times but would be restricted today. A simple example is arsenic, which although still commonly used in small amounts, was previously used much more extensively. The same is true for lead, which was commonly used for water pipes and in paint. Both of these elements, in various combinations, may be found in soil on which humans have built towns and businesses. Thus these chemicals may be present in appreciable amounts even if there is no indication that someone is living on or using the land at the present time.

It is also not safe to assume that contamination must be only at the surface or at a certain depth and no deeper. Material becomes buried for all kinds of reasons and by all kinds of activities, and so all sources of information about an area must be investigated, including historical accounts of activities, such as battles and settlements, that may have been located in the area.

11.2. AREA GEOLOGY

Knowledge of the geology of an area can be important in developing and carrying out a successful sampling plan and avoiding traps and mistakes. Two dramatic examples of this would be volcanic ash and karst landscapes. Both of these situations are unusual and can dramatically affect a sampling plan. Volcanic material may contain high levels of gravel-sized particles that facilitate movement of contaminants under saturated flow conditions. These soils often have very small amounts of clay, and this also leads to rapid movement of contaminants.

Karst landscapes develop in limestone rock and result in surface sinkholes and subsurface cave systems. In this type of landscape water, soil, and contaminants can move both on the surface and below it. Material can move to the underground cave system and then long distances until it emerges in a well, stream, river, or lake. Samplers need to be aware of the geology of the area in which the sampling is taking place and take into account the movement of material in different geological localities when developing a sampling plan.

It might be assumed that water and pollutants cannot move through solid rock, but exceptions exist. One is sandstone, which can be very porous and allow water to pass through it as rapidly as it passes through soil or sand. The second is the fact that in many cases underlying rock is not solid but fractured; that is, it contains many cracks through which water and pollutants can readily pass. Not knowing the type of underlying rock and its condition can lead to misunderstanding the fate of a component or contaminant in the environment.

In certain localities it is possible to have both soil and water move long distances. Soil can erode from sloping fields, in which case the amount of erosion and the final deposition of the eroded soil will affect the sampling plan. In addition to the erosion, soil can move downhill under the pull of gravity. This is termed colluvial movement and can result in the movement of large amounts of soil and regolith or underlying glacial till down a hillside, further resulting in significant movement of contamination if present. Both of these potential sources of movement must be accounted for in a sampling plan. If they are not, it is likely that not all the contamination will be discovered.

Table 11.2 shows soil analytical results that would be typical for a farm in central Ohio. At first glance, Table 11.2 does not look unusual except that sample WC-15c has a pH of 7.4, while all the other samples have acid pHs. One might then question the validity of this pH measurement. Statistical analysis would say that this is an outlier and could be discarded. However, knowing the geology of this area one would know that the regolith is basic while the overlying soil is acidic. This indicates that the sample is from an area in the field that has suffered severe erosion. An inspection of the field would show that this is indeed the case. Not knowing the geology of an area one would not be aware of this, and thus accurate data would be discarded without regard to its importance to sampling and analysis.

Another area of concern is water movement. Water can move rapidly over the surface of soil and carry contaminants into nearby streams, rivers,

TABLE 11.2 Soil Analysis Results for a Field in Central Ohio

Soil sample number	1:1 pH	cmol$_c$/kg CEC	kg/ha Ca	Mg	P	K
WC-15a	5.6	8	1,200	230	15	200
WC-15b	6.1	10	1,100	340	22	195
WC-15c	7.4	10	1,500	550	10	156
WC-15d	5.9	9	1,000	320	23	250
WC-15e	6.2	9	1,100	360	25	245

lakes, and other bodies of water or onto uncontaminated land. It is also possible to have water moving underground for some distance only to emerge aboveground at some other locality. All of the area in which the water is moving both below- and aboveground is important to sample and monitor. In addition to water carrying contaminants out of an area it can also carry contaminants into an area. Recognizing the potential for movement of contaminates out of and into an area will greatly simplify sampling.

11.3. KNOWING WHAT IS OR IS LIKELY TO BE PRESENT

For both safety and sampling it is essential to have some idea of the type and amount of contamination that is known or suspected to be present. The type and amount of protective equipment and clothing needed by sampling personnel will be dictated by this information. Relatively innocuous contaminants that do not produce hazardous fumes and are not toxic will require minimal safety equipment and precautions. On the other hand, high levels of contamination and contamination with highly toxic materials will require the maximum in safety equipment and precautions.

The type of sampler and sample container used will be determined by the type of contaminant thought to be present. Samples thought to contain volatile organic compounds (VOCs) need to be sampled and placed in sealed containers without headspace.* This is necessary to prevent the loss of VOCs before the analysis can be completed. On the other hand, soil to be analyzed for metals need not be stored in containers that are airtight. However, in this case they must be kept away from metal contamination,

---

* Headspace is the space (air space) between the sample and the top of the container closure.

which might occur if they were stored in metal sample containers. Also, samples stored for metal analysis may lose other contaminants (e.g., VOCs), and thus borrowing a portion of this sample for some other analysis is not possible.

11.4. NOT ACCOUNTING FOR INPUTS, LOSSES, AND MOVEMENT

The level of a contaminant in water or soil will depend on the amount added to it, the losses, and its movement through or out of the field. Although this seems simple and it is, it is not the total story. It may be that the addition of a contaminant is a single, one-time event. On the other hand, there may be a continuing addition of contamination. This can be very confusing if the occurrence, source, and amounts of contamination being added are not known.

Additions can be made by spillage, but may also be due to water invading the area or by windborne material. A spill is likely to be known and may indeed be the reason for sampling. However, if additional material is being brought into the area by water or wind movement, it will make the sampling seem bad. Samples may show a continuing increase in contamination when there is no additional spill. The movement of water onto an area may be evident; however, the fact that it is carrying contaminant may not. On the other hand, wind may not seem a possible source of contamination when, through the deposition of airborne dust, it may well be. Analysis of water entering and air over a field may be essential in understanding what is happening.

Losses of contaminant through percolation, leaching, and erosion may be evident and measurable. Certainly there will be monitoring wells around the contaminated field, and from the analysis of the water in these wells the movement of contaminant out of the field will be known. However, erosion of contaminated soil may not be as easily identified. A hectare furrow slice of soil weighs approximately 2,000,000 kg, and thus the loss of 500 kg of soil will not be observable. For this reason, losses from this source may go unnoticed. Likewise, it may be that a small but significant amount of contaminant is escaping into the atmosphere. This also may be missed.

Mass movement of soil (as in sliding down a hill) can cause movement of associated contamination. This may mean that the area originally sampled no longer contains the contaminant, thus requiring samples to be collected downhill from the original sampling area. Sampling thus needs to follow the direction of the movement of the contaminated medium. Failure to take this into account can result in an area being declared free of contamination when all that has happened is that the contamination has moved to a new location. This can come back to haunt both samplers and remediators.

11.5. PERSONNEL

Businesses will always prefer to hire the least expensive labor they can. Because field sampling appears to be simple it seems reasonable to hire samplers who have little education and no background in field sampling. This can and will lead to sampling inconsistencies that will cause severe and sometimes costly or even deadly mistakes.

These mistakes will be made in six areas. The first and simplest is that the correct field will not be found or the incorrect field will be sampled. The second set of mistakes will occur in reading maps and judging distances. The third will occur because representative, consistent, uniform samples will not be taken. Observing of environmental characteristics, maintaining sample integrity, and observing safety rules will be three more problems associated with inexperienced samplers. It is hoped that only one of these is the major problem and can easily be rectified. Unfortunately, several of these may occur in the same sampling event.

For the inexperienced samplers problems start with not finding the correct field, and it is often the case that inexperienced samplers will sample the wrong field. However, this can also happen with trained personnel if they are not careful. Often fields and their extent are not easily recognizable. They will not have GPS references associated with them. Personnel must rely on tree lines, streams, roads, buildings, and their relationship to each other to find the correct field.

Assuming the person finds the correct field, how is he or she to set up the sampling grid? Experienced samplers will be able to accurately judge a grid in a field by picking out reference points around the field. Inexperienced samplers will take all their samples from a relatively small, unrepresentative area in a large field. If multiple grids are to be sampled and multiple subsamples taken from each grid, then such persons will fail to sample all the grids appropriately. Upon completion of sampling the position of the samples in the grided area will be unknown.

If samples are not taken from representative places in the field, are taken to different depths at different places, or are not taken in the prescribed place in the field, the analysis of these samples will be accurate but may not be representative of the field. These eventualities will come about because areas are hard to get to, it is harder to insert the sampler in some places than in others, and some designated places in the field will not be found.

Another concern is that samplers understand environmental factors affecting sampling. Often variations in layers and direction of movement of the medium being sampled are not taken into account. This leads to mixing of samples from different layers in one sample container. Other mistakes would be not noting changes in layer thickness or density in the project notebook and not taking these into consideration in obtaining the sample. For instance, if the A horizon is specifically being sampled and it becomes thin in an area, it is not acceptable to include the B horizon in the sample just to make it the same depth of other sample cores. In this case, this variation needs to be detailed in the project notebook so that later evaluation of variations in the A horizon and how they affect sampling and analytical results can be made.

The inexperienced sampler will often not be careful in maintaining the integrity of the samples. This may mean that the sampling tool is not cleaned between samples, that care is not taken in getting all the sample in the container, or that the container is not sealed or is not placed in the proper transportation box. It might also mean not taking care to protect the sample from heat, cold, light, and contamination after sampling or not transporting it rapidly to the analytical laboratory. An additional problem with the inexperienced sampler is in not being careful to make sure that all data are carefully recorded in the project notebook. Careful recording of field notes is essential in later interpretation of analytical results and applying them to the remediation of a field problem. These notes are also essential in deciding if data are correct and if extraneous or suspicious data need to be kept or discarded. This is especially important during statistical analysis of analytical results.

11.5.1. Using GPS

A particularly effective way to solve many of these problems is with the use of GPS. Fields can be unequivocally located using GPS, and waypoints or sampling sites can be located and their positions programmed into the GPS instrument being used. In this way the location of the field and sample sites therein are known and the sampler can locate them by referring to the GPS unit. Both experienced samplers and inexperienced samplers will need training in reading, interpreting, and recording GPS data. It is essential that the person sampling be able to accurately read and record GPS data in the project notebook. This is particularly important if the sample is not taken at the exact location specified but nearby, or if the field is to be repeatedly sampled. While learning to use a GPS instrument it is important that samplers practice sampling under the supervision of an experienced sampler.

It might be thought that having the sampler physically mark the sampled sites would solve this problem—it does not. With inexperienced samplers, checking sampling markers placed at the sampling sites will show that the field was not sampled correctly. The solution to this problem would be for the field to be resampled correctly. Time is an important component of sampling, and resampling may not be possible for several weeks after the original sampling event because of weather or work schedules. Thus, the original samples will not be comparable to the later samples. For some situations it may be critical for the success of the sampling and remediation process that samples be taken at a particular time. In this case resampling can have a large detrimental impact on the whole sampling project [2]. Common mistakes made by inexperienced samplers are summarized in Table 11.3.

11.5.2. Safety

Safety will also be of special concern, especially for inexperienced field samplers, and the first concern is for the safety of the person doing the sampling. This is due to the fact that some samples or fields can be dangerous to human health, and this fact may only be discovered after analysis is complete. All workers must observe all safety rules at all times. In addition to this being important for the person doing the sampling, it is also important for persons helping with the sampling. The second concern is for the safety of sampling equipment and the samples already taken. In the sampling process money and time is expended in purchasing and maintaining sampling tools and in obtaining the samples.

TABLE 11.3 Mistakes Made by the Inexperienced Sampler

Field	Not finding the field and the corresponding field map
Sampling	Taking samples from the incorrect places in the field
Sample size	Not taking samples from the same depth at each site
Horizons	Not noting changes in horizons and direction of movement
Project notebook	Not recording all observations in the field notebook
Safety	Two aspects
	1. Not following safety rules
	2. Not operating and maintaining equipment safely
Sample handling	Not handling samples in the prescribed manner

11.6. ANALYZING OR SAMPLING FOR THE WRONG COMPONENT

An ammonia spill occurs in an open field. The weather and soil are warm and there is a gentle misting rain each evening keeping the soil moistened at field capacity. Several weeks after the spill, soil and water samples are taken. Ammonia levels are found to be at acceptable levels and so the assumption is made that there should be no concern about the spilled ammonia. It is obvious that much less ammonia has been spilled than was reported. A week after the sampling, analysis, and declaration that no problem exists, several babies in nearby houses that use well water for drinking come down with blue baby syndrome [3]. Could there be a connection?

Under the soil and environmental conditions described above ammonia is rapidly oxidized, first to nitrite and then to nitrate, which subsequently leaches into the groundwater. Under the conditions described soil bacteria can convert kg of ammonia to nitrate per day. Thus in a very short period of time the level of ammonia is back to its background level. In this case analysis should be not only for ammonia, but also for nitrate.

A company is accused of dumping chromate, specifically chromate(VI), which has many adverse health affects, on a field. Soil analysis shows only the occurrence of small amounts of Cr(VI). In this case it is important to know that Cr(VI) is quickly converted to Cr(III) in soil, and so this is the species for which the soil should be analyzed.

The types of changes illustrated above can also occur with organic contamination. For example, a monitoring well at a landfill that had been closed and capped years before was observed to have rapidly increasing levels of vinyl chloride, which had some other contaminants associated with it. It was hypothesized that illegally dumped or lost drums of vinly chloride were the source of this chemical in the wells. However, all compounds were found to be chemical and biological decomposition products of solvent mixtures in decomposing drums 10 ft deep in the landfill.

11.7. ANTAGONISMS AND INTERFERENCES

There are numerous antagonisms between analytes and between analytes and matrices. For instance, it is possible for two different compounds to elute from a gas chromatograph column at the same time, thus obscuring the occurrence of both compounds. Often a matrix will mask the occurrence of an analyte, making it appear not to be present. The analytical laboratory can change analytical and extraction procedures to take into account different retention times or other confusing interactions with analytical methods so that an accurate measurement of the analyte can be made.

These types of method adjustments cannot be made without the laboratory having a complete history of the field as well as the nature and levels of contamination present. However, in some cases, well-meaning but sometimes not well thought out actions of the persons submitting a sample will result in interferences. A good example of this is a field that was used as a wood treatment facility and was heavily contaminated with both organic and inorganic contaminants. Soil cleanup was being done by arc furnace, which reduced the soil to ash. The ash was analyzed for such compounds as benzene, naphthalene, polyaromatic hydrocarbons (PAHs), and pyrenes to ascertain if they were within regulatory limits. Purge and trap was used for benzene, the other components were analyzed by extraction, and initially none of these compounds was found in the ash.

After 2 months and hundreds of samples benzene started to be observed in samples. However, its recovery was poor and the observed concentrations inconsistent, varying from sub-ppb to 50–150 ppb levels. Little change happened during the following 2 weeks of checking and further testing except that the recovery levels got worse and more erratic. During this time the client asserted that no change had been made in its sampling procedures and so the problem had to be in the laboratory procedures. After another 2 weeks the client admitted that it had started adding lime to the samples before incineration.

It was determined that the poor recoveries and inconsistent analytical results were attributable to this lime addition. During incineration lime was reduced and when mixed with the ash produced cement. Thus when water was added to the samples during extraction the contents were solidifying and this interfered with the purge and trap process. This explained the erratic results, but not the occurrence of benzene in the sample because it should have been destroyed during incineration. At this point the client again said that nothing had changed in its procedures that would account for the appearance of benzene.

It turned out that the client had begun using water to cool the ash when it exited the furnace. The cooling water was being taken from an open pit at the field, and was contaminated with various chemicals leached from the contaminated field's soil and into the water in the pit. Benzene was the most mobile of these chemicals and so was the first to show up during analysis. Water in the pit was found to contain benzene in excess of several hundred ppb. The good intention of cooling the waste ash to speed the process resulted in several hundred tons of ash and soil being passed through the furnace, again at considerable expense.

This illustrates the incompatibility of procedures being used at the sampling field with those being used in the analytical laboratory. Note here that a detailed description in a project notebook of procedures being carried out in the field would have greatly facilitated finding the source of the problem and correcting it and would have saved considerable time and money.

11.8. SAMPLE ANALYSIS MISMATCH

Two of the most common mistakes in the commercial laboratory are not considering either all the aspects of the sample that is being analyzed or the basic question that is being asked; that is, can the sample be tested for the analyte in question or will the sample preparation or analysis introduce too many problems? An example would be looking for VOCs in a solid sample that needs to be ground before sample extraction. The extra sample preparation and handling required will result in VOC loss unless steps are taken to prevent it. The sample may have to be handled under an inert gas, at a reduced temperature, or by manipulation within a sealed environment. The resulting headspace as well as the original sample then needs to be analyzed.

Another example of a sample and test mismatch would be the extraction of a plastic or resin polymer sample with an organic solvent that could partially or fully dissolve the underlying polymer base. A sample that contains plastic fragments may have some plastic dissolve while it is being extracted using a large volume of methylene chloride, and this may interfere with subsequent analysis. Some of the dissolved plastic will precipitate when using solvent evaporation methods to concentrate the solution to the final analytical volume. This may result in a sample being impossible to analyze because either the whole sample solidified or part of the component of interest is dissolved in or sorbed by the precipitated plastic.

11.9. EXTRACTION OR METHOD ERRORS

Analyzing a sample for a target analyte may require solvent extraction and subsequent concentration. At this point in the analysis the concentrated solvent may be incompatible with the instrument or detector stipulated in the original method. As an example, a water sample analyzed for pesticides is extracted using diethyl ether and is then concentrated and dissolved in methyl tertiary butyl ether (MTBE). This concentrated extract is then analyzed using a gas chromatograph and an electron capture detector. For ether extracts and solvents detector response is acceptable and there is no interference with target compounds.

If on the other hand an equally common extracting solvent, methylene chloride, is used in the extraction and subsequent concentration, an acceptable detector response will not be obtained with this prescribed detector. The extraction and concentration steps will proceed well and the final concentrated analytical sample will look good. Indeed, it may even look like the MTBE extract described above. However, when the sample is injected into the gas chromatograph the electron capture detector, which is not compatible with this solvent, will become saturated with methylene chloride, requiring several hours of cleaning before it can again be used for the desired analysis.

11.10. SAMPLE/TEST CONTAINER ERRORS

Sampling one type of matrix using the wrong type of container or preservative can have dramatic and catastrophic effects on the data obtained from the sample. An example of this is the use of plastic containers to transport and store samples to be analyzed for organic contaminants. Plastics commonly contain phthalates, which are a class of organic compounds used to give a plastic flexibility. Small amounts of these compounds can readily leach out of plastics and interfere with subsequent analytical procedures. Phthalates could also be "extracted" by the organic compounds in the sample being analyzed.

Samples can also be contaminated by using a batch of sample containers that are not properly cleaned and by not cleaning sampling equipment thoroughly between uses.

11.11. IMPROPER SAMPLING TECHNIQUES

Inattention to sampling methods can lead to errors in sample analysis. An example would be leaving headspace in a sampling container intended for VOC analysis. Sample components will volatilize into the headspace at different rates, depending on their vapor pressures. This leads to an uncharacteristic sample and erroneous data, which can lead to inappropriate actions being taken in the cleanup of the field.

When sampling soils, great care needs to be taken to ensure that the threads, the lid of the container, and the seals are clean and free of particles. Such particles can prevent proper sealing and subsequent loss of sample through vapor loss, evaporation, or leakage.

It is imperative that care be taken to make sure that a representative sample is collected. This point is especially important for soil or other solid samples, especially when distinct layers are present. It is equally important in situations in which liquid samples are made up of several immiscible layers. Failure to note this and ensure that a representative sample or samples comprising all the fractions or layers is taken will result in erroneous data and analytical results.

One way to handle this problem is to sample each of the distinct layers or fractions individually and separately. Each of these fractions would be analyzed using the procedure appropriate to either the type of material or the contaminants it contains. The resultant analytical data can be put together to produce a composite picture of the contamination in the field; for example, if the samples are taken from a pile of excavated soil, the outside of which may be lower in contaminant because of exposure to light and air. In this case samples can be taken immediately after excavation or from both surface and subsurface layers in the pile. The resulting analytical results can be combined to describe the concentration gradients in the pile.

11.12. VARIATIONS IN METHOD

Variations in extraction techniques can change analytical results and interpretation. This could be almost anything, such as using the wrong solvent or even using a different amount of solvent at some stage in the extraction. Variation in the time of extraction, not using prerinsed glassware, and not using prerinsed drying agent can also lead to sample contamination or loss. This is one reason for extraction surrogates. They provide a reference for extraction efficiency. Most extraction methods have set limits for the extraction efficiency of the surrogates as per the method. When the efficiency is within limits the data will be readily acceptable. If the solvent recoveries are outside acceptable limits the extraction is suspect and must be redone.

For clean and relatively low viscosity samples extracts will produce good results. Dirty or highly viscous samples are problematic and may produce erratic results. For a number of reasons potential problems from dirty or viscous samples should be considered before, during, and after analysis and review to determine the usefulness of the data.

11.13. MOVEMENT OF A CONTAMINANT

The movement of a contaminant is usually derived from the site's underlying hydrology. The water table will move the contaminant plume mainly in the direction of the water flow. This is especially true for those chemical compounds with very high solubility, for example, MTBE. For contaminants with low solubility in water the diffusion rate from the point of contamination will be much lower. Soil, water, and air sampling must always be carried out downstream from the point of contamination. This does not mean that sampling to the side, uphill, upstream, and upwind from the contamination is not necessary. It is. This goes along with mass movement of soil as noted above, which must also be kept in mind.

11.14. STATISTICAL ERRORS

In looking for sources of sampling errors most literature searches will provide a multitude of references that are primarily statistical analyses. These are books and papers that show how statistics can be used to detect such things as errors. However, the use of the incorrect statistical tool or method, including the incorrect use of hypothesis or method of calculation, the use of an incorrect table, or the incorrect interpretation of the statistical results in the analysis of data, are dangerous traps which must be avoided. To do this it is highly recommended that an experienced statistician be part of the investigatory team or be used in doing the statistical analysis on the data [4,5].

11.15. CONCLUSIONS

Many traps, mistakes, and errors can take place during sampling and analysis of environmental samples. Personnel and activities in the field are the two biggest sources of errors in sampling and analysis. Knowing the history of an area, its geology, what is present, and the rates of inputs, losses, and movement is also important to accurate sampling. In the analytical procedures, analysis for the wrong component, antagonisms between analytes, sample analysis mismatches, and extraction using the wrong method are important sources of errors. The characteristics of the environment and movement of contaminants in that environment are important considerations. The use of proper statistical tools is essential in obtaining good sampling results.

QUESTIONS

1. Describe the training personnel should have before beginning to field sample. Include all aspects of sampling.
2. Think of an example in which knowledge of the history of a field would be important in developing a sampling plan for that field.
3. A soil sample reported to be contaminated with lead is sampled and the analysis shows a lead level of 10 ppm. What would you need to know before deciding that the field from which the sample came needed to be remediated to remove lead?

4. A soil supposed to be contaminated with hydrocarbons is extracted with aqua regia (a strong acid). The result is a solution that cannot be analyzed for hydrocarbons. Explain why this is the case. What would aqua regia be expected to extract from a soil sample?

5. The results of analysis show that in week 1 the level of zinc in a field was 120 ppm. The sample from a week later showed a level of 140 ppm. It is assumed that the analysis is in error. What other explanation could there be for this increase, assuming it is not an error?

6. Iron (Fe) in soil reacts with phosphate to form insoluble precipitates. After the application of phosphate fertilizer to a field what would be expected to happen to the level of soluble iron in the soil?

7. A field is contaminated with large amounts of nitrate. A 0-to-20-cm surface sample from this field shows little nitrate. Does this mean that the nitrate has disappeared from the soil? Where would you sample to make sure the nitrate is not in the soil? If the nitrate is not in the soil, where might it be?

8. A sample extract used to remove VOCs from a sample is analyzed by atomic absorption. An absence of metals is seen in the results of the analysis. Can you explain why this might be the case?

9. A statistical analysis of the results of a series of extracts of the same soil shows a large variation in the results. Is this to be expected?

REFERENCES

1. Leschber R. Sampling: Fundamental aspects. In: Gomez A, Leschber R, L'Hermite P, eds. Sampling Problems for the Chemical Analysis of Sludge, Soils and Plants. New York: Elsevier Applied Science, 1986:2.

2. Gomez A, Leschber R, Colin F. Sampling techniques for sludge, soil and plants. In: Gomez A, Leschber R, L'Hermite P, eds. Sampling Problems for the Chemical Analysis of Sludge, Soils and Plants. New York: Elsevier Applied Science, 1986:80, 81.

3. Nitrates and Blue Baby Disease. Florida Department of Health. http://www.doh.state.fl.us/environment/water/SUPER_Act_website/watertox/nitrate.htm. 2003.

4. Myers JC. Geostistical Error Management: Quantifying Uncertainty for Environmental Sampling and Mapping. New York: Wiley, 1997.

5. Smith PL. A Primer for Sampling Solids, Liquids, and Gases: Based on the Seven Sampling Errors of Pierre Gy. Asa-Siam Series on Statistics and Applied Probability. Philadelphia: Society for Industrial & Applied Mathematics, 2001.

Appendix A

Abbreviations and Acronyms

A	Estimated soil loss using the USLE
afs	Acre furrow slice
AM/FM	Automated mapping/facilities management
AMU	Atomic mass unit
APHA	American Public Health Association
ATT	Abbreviations and acronyms and terms
AWWA	American Water Works Association
BMP	Best management practice(s)
CADD	Computer-aided design and drafting
CD	Compact disk
CD-R	CD-record only
CFC	Chlorofluorocarbon
CHP	Chemical hygiene plan
COC	Chain of custody
DAAC	Distributed active archive centers
DGPS	Differential global positioning system
DI	Deionized (usually water)
DO	Dissolved oxygen
DOQ	Digital orthophoto quadrangle
DOQQ	Digital orthophoto quarter quadrangle
DOT	Department of Transportation
DVD	Digital video disk

DVD-R	DVD-record only
EDD	Electronic data deliverable
EH&S (EHS)	Environmental health and safety
EOS	Earth observatory system
EPA	Environmental Protection Agency
EROS	Earth resources observation systems
F	F statistic
FAA	Federal Aviation Administration
GIS	Geographic information system
GMT	Greenwich mean time
GPO	Government Printing Office
GPR	Ground-penetrating radar
GPS	Global positioning system
hafs	Hectare furrow slice
HMIS	Hazardous material identification system
ISE	Ion selective electrode
K	Hydraulic conductivity
LD_{50}	The dose at which 50% of test animals die
LIMS	Laboratory information management system
MCL	Maximum contamination level
MDL	Minimum detection limit
MSDS	Material safety data sheets
MSE	Mean square for error
MST	Mean square for treatments
NAD-27	North American Datum-1927
NAD-83	North American Datum-1983
NAPL	Nonaqueous phase liquid
NASA	National Aeronautics and Space Administration
O_3	Ozone
OSHA	Occupational Safety and Health Administration
PAH	Polyaromatic hydrocarbons
PCB	Polychlorinated biphenyls
PM	Project manager
PM10	Particulate matter $10 \, \mu m$ or smaller
ppa	Parts per quintillion
ppb	Parts per billion
PPE	Personal protective equipment
ppm	Parts per million
ppq	Parts per quadrillion
ppt	Parts per trillion
RCRA	Resource Conservation and Recovery Act
RFA	Request for analysis
RUSLE	Revised universal soil loss equation
RWEQ	Revised wind erosion equation
s	Standard deviation

s^2	Sample variance
SCF	Supercritical fluid (extraction)
SPE	Solid phase extraction
SSE	Sum of squares for error
SST	Sum of squares for treatment
STP	Standard temperature and pressure
t	t-statistic
TCP	Transport control plan
theta	Effective porosities
TPH	Total petroleum hydrocarbon
TOPO	Refers to topographical map
USB	Universal serial bus
USEPA	United States Environmental Protection Agency
USLE	Universal soil loss equation
UTC	Universal time coordinate
UTM	Universal transverse mercator
VOC	Volatile organic compound
WAAS	Wide area augmentation system
WEF	Water Environment Federation
WEQ	Wind erosion equation

Appendix B

Sources

SAMPLERS

Clements Associates Inc.
1992 Hunter Avenue
Newton, IA 50208

Giddings Machine Company, Inc.
401 Pine Street
P.O. Drawer 2024
Fort Collins, CO 80522

MODELING

Rock Ware
2221 East Street
Suite 101
Golden, CO 80401

ERSI
380 New York Street
Redlands, CA 92373-8100

GENERAL SUPPLIES AND CONTAINERS

Eijelkamp
P.O. Box 4, 6987 ZG Giesbeek
Mijverheidsstraat 30
6987 EM Giesbeek
The Netherlands

Gempler's
100 Country Side Drive
P.O. Box 270
Bellsville, WI 53508

Spectrum Laboratory Products
14422 South San Pedro Street
Gardena, CA 90248-2027

SAMPLE ANALYSIS
Elabs, Inc.
8 East Tower Circle
Ormond Beach, FL 32174

SAFETY
Ullom Safety Resources, Inc.
P.O. Box 340601
Beaver Creek, OH 45434-0601

REFERENCE SOURCES FOR SAMPLING METHODS AND SAMPLE ANALYSIS
USEPA—http://www.epa.gov/epaoswer/hazwaste/test/main
ASTM—http://www.astm.org
APHA—http://www.apha.org/media/
AWWA—http://www.awwa.org/
WEF—http://www.wef.org/
OSHA—http://www.osha.gov/
ASTM—http://www.astm.org/

Index